THE UNDERWATER NATURALIST

THE
UNDERWATER
NATURALIST

David K. Bulloch

AN AMERICAN LITTORAL SOCIETY BOOK

Lyons & Burford, Publishers

TO EDITH

8 circ 11/21/98 (9/99)

Printed in the United States of America

10 9 8 7 6 5 4 3 2 1

Designed by Ruth Kolbert

Typesetting by Fisher Composition, Inc.

Library of Congress Cataloging-in-Publication Data

Bulloch, David K.
    The underwater naturalist / David K. Bulloch.
      p.   cm.
    "An American Littoral Society book."
    Includes bibliographical references and index.
    ISBN 1-55821-108-X
    1. Marine biology. 2. Man—Influence on nature.
I. Title.
QH91.B78   1991
574.92—dc20   91-4690
                                CIP

Contents

Acknowledgments

This book has been written in part to repay a debt that I owe to those who long ago helped me learn about life in the sea. As with most debts incurred early in life they can no longer be resolved directly but can only be settled by passing along the legacy.

My thanks to Derek Bennett, Executive Director of the American Littoral Society for access to the Society's library, its files and photographs. I have also drawn from contributions to UNDERWATER NATURALIST, the Society's quarterly publication.

Members of the National Marine Fisheries Service at Sandy Hook, NJ have been most helpful especially Claire Steimle and Judy Berrian for guidance into the marine literature.

Shared sea experiences with Charlie Stratton, C. Douglas Hardy, Bill Jahoda, and other friends have each contributed in their own way.

And to Edith for help, patience, and love over many years.

1

Life in the Sea

WHY THE SEA ATTRACTS US IS BEYOND RECKONING. BUT IT does, powerfully, first drawing us to its shores, then tempting us out on its wide expanses, and in recent years, urging us to explore beneath its surface.

The sea is a beckoning wilderness. Beneath its waves curious creatures do mysterious things. We have discovered much about some of them, but even the lives of the most common remain relatively unknown. Few people recognize more than a handful of the sea's inhabitants; fewer still have the slightest notion how they live. And for good reason. The oceans are not open to ready view. The sea hides its progeny under a watery veil. We see only what washes ashore with the tide or what has been hauled up by hook, dredge, or trawl.

Just a few miles off the shores of the United States, the waters are as foreign to us as the heart of Africa or the farthest reaches of the poles: a realm barely explored, a wilderness we have yet to comprehend fully. Its deeps can only be fathomed by instruments or submersibles, but its shallows lie within your reach. With mask, fins, and snorkel you can go into the water and see for yourself what is below the surface. With training you can explore deeper on self-contained underwater breathing apparatus (SCUBA).

Go into the water and look below the surface. What are all those creatures you see? You recognize fish and a few other forms easily enough, but what about all these others; what are they called and how do they live?

On land, you recognize nearly every living thing you lay eyes on. You know what it is, what it does, and whether or not it is behaving normally. Not so in the sea. Initially, you may not be able to tell an animal from a plant. Unlike those on land, sea animals are untroubled by gravity or desiccation and have taken what are to us odd shapes and habits. Some have found that staying put is easier than crawling, swimming, or floating.

Because all the world's seas are contiguous, you might reason that all the sea's creatures are distributed worldwide. A few are, but most are not. Much of the life you will see is unique to the region in which you see it. Only a fraction of the animals and plants of the sea know no boundaries. Most are geographically limited, held in place by water temperature, bottom type, and other restraints. Even though similar underwater habitats abound worldwide, sheer distance and unhospitable environments between them isolate many similar kinds of life. Thus separated, over time they have evolved into distinctly different species.

Thus, each region has its own unique inhabitants living within their own particular habitats, be it sand flats, grassy beds, mud, or rocky bottom. The kind of bottom partly dictates the type of life that can live there: for example, clingers, diggers, or burrowers. Also important are the saltiness of the water, changes in the temperature of the water over the seasons, and water quality.

Years ago, water quality rarely hindered the distribution and abundance of plants and animals living near shorelines, but, today, our coastal waters have deteriorated so badly that once-common species are now rare or missing altogether from their former haunts. For example, oysters no longer survive in New York Bay. Habitats have vanished; for example, sea grasses have virtually disappeared from Chesapeake Bay. Heavily polluted harbors have almost barren bottoms; oil seepage from refinery operations has eliminated all but a single species of marine worm from Los Angeles harbor.

As you explore new places, you will continually uncover animals and plants new to you, and, initially, be at a loss to name them. Fortunately, you can tell most of the common ones apart by their external appearances alone. You may not be able to identify them precisely as to species, but you can broadly classify most into classes, families, or genera by sight. To distinguish species you will need keys: guides

Getting below is half the fun. An experienced free diver can not only explore to a depth of twenty feet with ease but can remain long enough to photograph, retrieve specimens, and make general observations.

that point out differences and similarities among the bewildering array of forms you will uncover. This is especially true for the invertebrates, animals without backbones, which are so diverse and so common in the sea, yet so foreign to us on land.

You can find field guides to each major shoreline region of the United States that cover the common fishes, mammals, and invertebrates found there. They are arranged much like guides to land creatures: things with similar shapes are grouped together. Where possible, select narrow regional guides in preference to broad, all-inclusive ones. The comprehensive ones will overwhelm you with too many possibilities, which you must sort out by noting very small differences. Until you are well acquainted with a large number of the members of the basic groups of animals, broad coverage can be more confusing than helpful.

When it comes to distinguishing between closely related species, especially those not well known or popular, you may find even regional guides too vague and will have to seek more specific help. No naturalist, or professional biologist for that matter, can hope to know all the species in a region. They must rely on texts and monographs written by specialists who have studied the systematics of specific groups.

Identifying a creature is a far cry from knowing it. Where it lives, what it does, and creatures it associates with are fully as consequential as a description of its anatomy and its name. The life histories of only a relative handful of all known animals have been worked out to date. It is here that the amateur naturalist can make significant contributions to learning. Here the patient amateur, unhurried by the pressures of the professional, can uncover changes in patterns of living over seasons, over years, as animals and plants respond to their environments.

GEOGRAPHIC DIFFERENCES

The sea, like the land, has its own topography and physical settings that influence the distribution of living things within it. Some species of plants and animals range oceanwide; others are narrowly confined. Regions physically similar but widely separated harbor similar kinds

Major marine provinces of the U.S. coastline. The Carolinian and Louisia-nian biomes are alike because they were once connected when the Florida peninsula was submerged.

of life, but invariably they are wholly different species, the result of isolation as they evolved. Wave action, currents, tides, dissolved oxygen, salinity, nutrients, water temperature, bottom type, and seasonal changes all set the stage for what life forms live where.

Biologists have divided the coastlines of the continental United States into seven regions, or provinces. Some species overlap at the boundaries of these provinces either seasonally or permanently, but water temperature and bottom type confine most of the creatures to their own appointed places.

Our Northeast Coast, from Maine to Cape Cod, lies in the *Acadian* province. The coast is rocky and the water is chilled by the Labrador Current. Summer surface water temperatures hover near fifty degrees Fahrenheit (F). Large tides, averaging nine to eighteen feet, create swift coastal currents. Rockweeds and kelps cover the shallow bottoms, and much of the animal life found there has adapted to clinging either to rocks or weed. Fewer kinds of fish live in these cold waters than in the tropics, but those that do multiply rapidly, providing some of the richest fishing grounds in the world: Georges Bank is among the best-known of them.

From Cape Cod to Cape Hatteras, North Carolina, lies the more temperate *Virginian* province. Although its waters during the winter can be as cold as those of the Acadian province, they are far warmer in summer, with temperatures rising into the seventies. Tides are more moderate, averaging from one to four feet. The bottom is sandy with only a few scattered rock outcrops. Without hard surfaces, burrowers dominate here. Extensive beds of sea clams, for instance, prosper in the sandy bottoms.

Below Cape Hatteras to Cape Kennedy, Florida, the waters are warmed by the Gulf Stream. Extensive marshes line the shores, protected from open ocean waves by barrier islands. The life in this *Carolinian* province is similar to the *Louisianian* province, which stretches along the northern coast of the Gulf of Mexico, reflecting a time when Florida was completely submerged by the sea and those coastlines lay adjacent to each other. Winter water temperatures rarely fall below 50° F. Tides in the Gulf are small, averaging 1 to 2 feet, while those along the lower East Coast vary between 2.5 to 7 feet.

Below Cape Kennedy, on the eastern shore of Florida to Cedar Key on its western side, and southward from Port Aransas, Texas, to the southern portion of the Gulf, the Yucatan Peninsula, Central America, and the Caribbean Islands, lies the *West Indian* province. Winter water temperatures seldom fall below 70° F. Corals and mangrove proliferate. Tides are small, 1 to 2.5 feet.

The continental shelves of the East and Gulf coasts are broad, in places hundreds of miles wide. The shelf along the West Coast is narrow, more in the order of tens of miles and, in a few places, virtually nonexistent.

The West Coast of the United States, excluding Alaska, is divided into two provinces. The *Californian* province extends from Mexican waters north to Cape Mendocino, California. Although sandy to the south, most of the California coast is rocky. Freshwater runoff is low. The California Current creates a Mediterranean climate on land, but the waters are cool enough to support great kelp beds along much of the coast. Winter surface-water temperatures range from 50° to 55° F, and summer temperatures 70° to 74°. Tides are moderate, three to four feet.

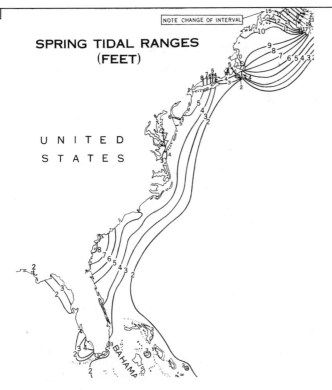

SPRING TIDAL RANGES
(FEET)

NOTE CHANGE OF INTERVAL

U N I T E D
S T A T E S

BAHAMA

EBB AND FLOW

Large tides have two practical consequences for the underwater naturalist: They significantly alter the distance from the water's surface to the bottom; and, depending on the configuration of the coast, can produce swift currents.

On Cape Cod, a low spring tide can reduce the water depth by three meters and a full running ebb, or flood tide, in a narrow channel can flow so swiftly it will submerge anchored navigation buoys.

Tidal cycles result from the gravitational interaction between the earth, moon, and sun; the continually changing positional relationships among them as they orbit around one another; and the earth's daily rotation.

The rotation of the earth and the motion of the moon create a tidal oscillation from low to high every twelve hours and twenty-five minutes. Tidal range peaks during the full moon and new moon *(spring tides)*, and is at a minimum during the first and last quarters *(neap tides)*. This cycle, a lunar month, takes twenty nine and a half days.

Because the moon's declination varies from 28.5 degrees north to 28.5 degrees south over a lunar month, the range of the two daily highs and lows not only change, but are not equal. Combined with changes created by the natural period of oscillation of a body of water that, in part, is governed by the configuration of the basin in which it rocks to and fro, at a given place daily tides may be *semidiurnal, diurnal,* or *mixed*.

Semidiurnal tides, two highs and lows every twenty-four hours and fifty minutes—one lunar day—are about equal heights. A *diurnal* tide has but one high and one low each lunar day. *Mixed* tides have two highs and two lows daily, but of unequal height. Tides on the East Coast are semidiurnal, and on the West Coast are mixed.

From Cape Mendocino northward to Vancouver Island, British Columbia, lies the *Columbian* province, whose foreshores and headlands are steep and rocky. Although tempered by water from the south, the cold Aleutian Current brings with it the life of the far north. Winter water temperatures range between 49° and 54° F, and in summer rise only a few degrees higher. The tidal range is higher here than to the south, averaging 4.5 to 7.5 feet.

Each province has its own unique species as well as overlapping creatures from nearby regions. Yet the provinces have similarities in habitats that allow us to group them—open sea, sandy bottom, rocky bottom, tropic shores, and man-made places—and look at what is common as well as what is unique about them.

THE WATER ITSELF

To the life in the sea, the water is everything. It brings food in the form of trace minerals, dissolved and suspended nutrients to those fixed in one place. It supplies oxygen, carries away metabolic waste, maintains internal salt balance, provides buoyancy, and moderates daily and seasonal temperature swings.

Temperature regulates the distribution of life in the sea more directly than any other single factor. Temperature differences between regions can create barriers that are as effective as underwater walls. Many animals and plants breed, grow, and do well only between narrow temperature ranges. Some can live only in warm water, others only in cold. For example, reef-building corals rarely proliferate in waters below 70°F. or above 95°F., doing best at about 82°F. The Portuguese man of war, a jellyfish, plies warm, tropical seas, whereas its cousin, the lion's mane, drifts exclusively in frigid northern currents. In temperate waters, whose winter-to-summer temperature variations may exceed forty degrees Fahrenheit, the life cycles of many fish and their migrations are closely attuned to these changes.

Less oxygen will dissolve in warm water than in cold; the amount is reduced by half between 32°F. and 85°F., thus trouble from high temperatures for some animals can be as much the result of suffocation as of overheating. Inshore, in summer, naturally low oxygen levels are often exacerbated by pollutants that consume oxygen as they degrade. Fish trapped or driven into such shallows by their enemies can perish from the effects of the depleted oxygen.

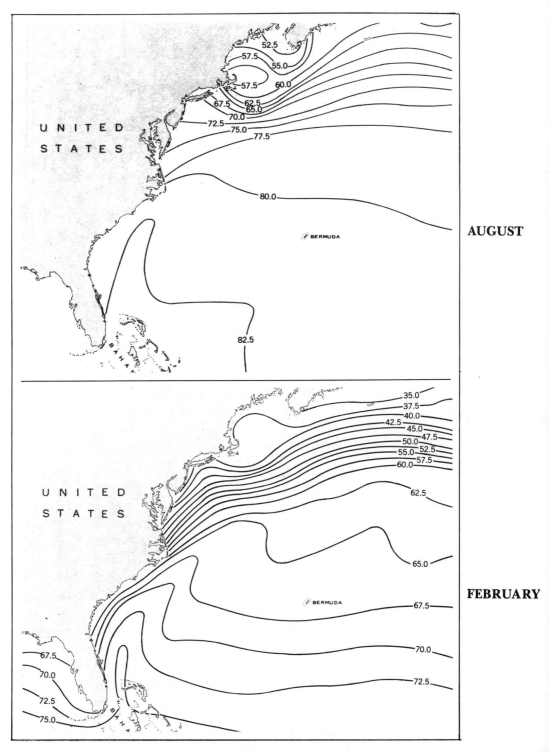

Winter and summer sea surface temperatures (Fahrenheit) in the western
North Atlantic *(Illustration courtesy of U.S. Hydrographic Office, H.O. Publication 128)*

Dissolved oxygen can range between zero and ten parts per million in seawater. An animal's needs depend on water temperature, its level of activity, and its species. In cool, temperate (45°–60° F) water, fish begin to show signs of stress when oxygen levels fall below two parts per million, and will succumb if they cannot find better circumstances. Bottom life suffers as well. Some, like clams and oysters, can close up and survive a few days at depleted levels, but no animal can go on too long without relief.

Very cold water, although oxygen-rich, reduces the activity of temperate-water invertebrates. They stop feeding and cease moving, becoming easy prey to predators better adapted to the cold. Storms can upheave burrowers who are too numbed to redig. Winter gales occasionally heap windrows of clams and scallops onto beaches, providing an easy feast for ever-hungry gulls.

In open ocean, the saltiness of the sea is remarkably constant. Sea salt is more than sodium chloride, the major component of common table salt. It is a mixture of forty-nine elements. The ratios of the major elements to one another remain nearly the same in all the seas of the world. But total salt concentration, called *salinity* by oceanographers, does vary from place to place. Open ocean salinity averages thirty-four parts per thousand, coastal waters twenty-six or so. In bays, estuaries, and river mouths, salinity drops as one proceeds upstream, depending on tide, rainfall, and runoff.

Salinity partly partitions the distribution of plants and animals along the coast. Turtle grass will not grow in waters below twenty-five parts per thousand. The seaweed *Enteromorpha,* common in northern splash pools, can withstand salinity oscillations between zero and saturated brine.

You may find it useful to keep track of dissolved oxygen, salinity, and water temperatures, especially if you want to follow changes in stressed waters; for example, a polluted estuary. Measuring water temperature is easy to do; just remember that if you take a sample from deep water, you must measure its temperature immediately, before it has an opportunity to warm up. Measuring dissolved oxygen can be done chemically with simple kits that are available from laboratory supply houses at reasonable cost. If you get involved in a study that will require hundreds upon hundreds of readings, there is an instrument that measures dissolved oxygen directly: the dissolved-oxygen meter.

Salinity can be measured with a salinometer or, more approximately, with a hydrometer. (See Chapter 10 for more details.)

CLASSIFYING LIFE

Unlike that on land, most of the easily seen life under the water are animals, not plants. Among the plants, only seaweeds and sea grasses stand out; the great bulk of photosynthesizing life at sea is microscopic and no longer classified with the plants.

Of the million and a half or so species that have been named taxonomically so far, about a million are animals. Around two hundred thousand are found in the sea. In spite of this seeming paucity, the animal life of the sea is far more diverse than that of the land, representing more than thirty-two major phyla. Most of the species found on land are insects, representatives of a single class within a phylum.

The strange nature and appearance of so many of the sea's creatures hinders easy acquaintance with them. Recognizing a whale, a shark, or a tropical fish can be done instantly by almost everyone. But with the invertebrates, the animals without backbones, the going gets tougher. Some we know well: As a group, the mollusks seem simple enough, they are clamlike and snaillike (until you discover that the octopus is also grouped among them). Starfish, brittle stars, and sea urchins also have a common ancestry and are collectively called *echinoderms*. Others are completely foreign to us: moss animals, combjellies, arrow worms, sea squirts, and salps, for example.

These major divisions of the animal kingdom—Mollusca, Echinodermata, and another thirty or so more—are called *phyla* (the singular is *phylum*). Many of them are found only in the sea.

Seldom seen by humans, thousands of sea species have no common name. Others, widely distributed and frequently encountered, often have different common names in different locales. The answer to the dilemma about what to call something so that it cannot be confused with something else is the scientific name: genus and species. It is latinized so that it transcends both language and custom. **Blue crab** will do in Baltimore, but *Callinectes sapidus* will do anywhere in the world.

Every distinguishable creature has been given a genus and species name whose roots lie in a Latin description. (Freely translated, *Callinectes sapidus* means "beautiful swimmer.") Note that the genus is always capitalized, the species never is.

Sometimes the latinized name is followed in parentheses by the name of the person who named it and the year it was described. If you see (L.) after the Genus-species, Carl Linnaeus first characterized it. A Swedish botanist, Linnaeus devised the binomial nomenclature system in the middle eighteenth century, and it has stayed with us ever since.

Scientific names can change over time as creatures are reclassified or declassified. Defining a species is not always simple, especially when the definers are confronted with lookalike populations that won't interbreed or groups that appear separate but *can* interbreed. Lumping, splitting, and regrouping still go on as new evidence dictates, often requiring revisions in scientific names.

Some groups of species are so hard to tell apart that writers will refer to them by genus only. For example, species of the alga *Chlorella* look identical and can only be differentiated by measuring respiration rates.

Between the species and the phylum to which it belongs, lies a series of intermediate classifications. Similar genera are grouped into families, families into orders, orders into classes, and classes into phyla. Each step up includes more and more species whose collective points of commonality grow less specific as the groupings grow more general.

Keep in mind that the classifications are all man-made. The animals themselves pay no attention to them. Evolution's backing and filling blurs distinctions, creating conflicts in grouping. Fresh evidence from new methods also adds to the puzzle. DNA sequences in genes can suggest biochemical likenesses and differences not always reflected in structural appearance.

The hallmark of life is diversity. Individuals within populations and populations within species differ one from another. Ten million species now exist. One thousand times as many are extinct. Yet all life is related, originating from a common ancestor. Therefore all species have a next of kin whose ancestry goes back to those common beginnings.

We now divide living things into five kingdoms: animal, plant, fungus, protoctist, and prokaryote. The first three you recognize, the latter two require some explanation.

No one knows how life began, but all the life we know uses DNA (deoxyribonucleic acid) as its basic reproductive unit, which argues for a single successful ancestor. In the distant past, by some unknown mechanism, elementary compounds united into larger molecules, which transferred and consumed energy, duplicated themselves, and become structurally unique: in other words, alive.

Trapping the energy of sunlight in a pigmented structure preceded the development of the cell nucleus. Today's prokaryotes, cells without definite nuclei or complex organelles (substructures devoted to specific functions), either get energy from chemical conversion, as do the bacteria, or from photosynthesis, as do simple blue-green algae. Collectively, these creatures make up the kingdom of the Prokaryotae.

Over time, a more complex unit evolved, the *eukaryotic* cell. It contains a nucleus to take charge of reproduction and cell regulation, and specialized organelles to handle excretion, energy transfer, and other duties. These cells evolved into both single-cell and multi-celled organisms.

From the eukaryotic cell emerged the animals, plants, fungi, and wide collection of microorganisms once partly grouped among animals (the protozoa), plants (the algae), and fungi (the water molds and slime molds), but now heaped in one catchall kingdom, the Protoctista. Most are single-celled. Given that limitation, they conduct the business of living with surprising variety. The protozoa and algae come in a multitude of forms, from the shapeless amoeba to the intricately shelled diatom.

Most of the ocean's photosynthesizers are *protoctists:* green algae, blue-green algae, and diatoms. Collectively, they are called *phytoplankton* and are the primary producers of the sea in much the same way that shrubs, grasses, and trees are on land.

Animals: Lower Invertebrates

We are all familiar with the sea's vertebrates: fishes, both bony and cartilaginous (sharks, skates, and rays); mammals (seals, walruses, whales); reptiles (turtles, sea snakes, and a few lizards); and one or two rare amphibians.

It is the *invertebrates,* animals without backbones, that contain so many strangers. In geological time, the invertebrates predate vertebrates by billions of years, but just *when* they arose still remains something of an enigma. The fossil record of the Precambrian era, which ended about 600 million years ago, is chock full of algal mats, but holds few clear signs of animal life. Perhaps early animal life was too jellylike to leave fossils. Then, in the Cambrian period that followed, the fossil record is filled with all the main branches of the invertebrates we know today and many others that became extinct before the Cambrian period ended.

PORIFERA: SPONGES

Multi-celled animals arose from the Protoctista or an ancestor of both the protoctists and the animals. The exact line of succession is a mystery. The sponges, whose cells specialize but do not coordinate in the same way as other invertebrates, appear to be a side issue in animal succession and have often been called an evolutionary dead end. Nonetheless, they have been successful enough to remain with us and are still going strong.

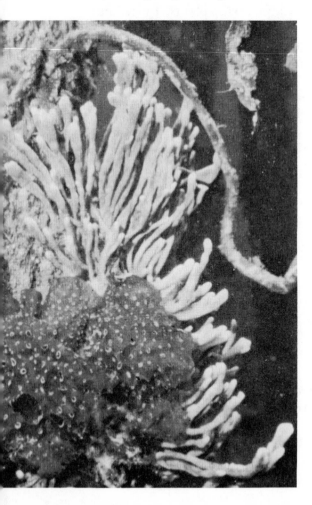

A typical North Atlantic Coast sponge, *Haliclona,* growing on a wharf piling *(Photo by Alan Stewart)*

All are *sessile,* that is, attached to something and unable to move on their own. Vase-shaped, encrusting, upright or fingerlike, their supporting framework is either a network or *calcareous* (calcium carbonate minerals) or *silica* spicules, or tough, interconnecting fibers. Their bodies enclose a series of canals and chambers through which seawater circulates in through pores in their walls and out through large openings. Flagellated collar cells whip the water and sustain its flow. A big sponge can pump bucketfuls of water and create a flow you can see with the help of a little ink. On your next visit to the tropics, take along some India ink in a squeeze bottle and squirt it deep in the center of a vase sponge. Watch the ink cloud rise in the current the sponge creates.

Sponges fall into three classes, only two of which inhabit shallow water: the Calcarea, who possess limy spicules, and the Demospongiae, who may have either silica spicules, a fibrous structure, or no stiffening material at all.

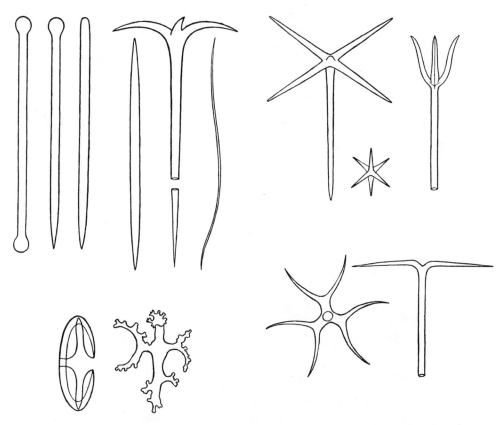

Sponge spicules. Certain sponges contain silica support rods and smaller links that can help identify them. To free these, carefully boil a bit of sponge in potassium hydroxide solution to dissolve the tissue, then examine the sediment under a microscope. *(Illustration courtesy of U.S. Fisheries Commission)*

CNIDARIA: JELLYFISHES AND KIN

Almost entirely marine, the jellyfishes are members of an early phylum that left no evidence of its beginnings in the ancient rocks, although corals, close kin, did. Those reef builders began their work a half billion years ago in the Ordovician period (which followed the Cambrian) and have been at it ever since.

Their body plan shows the beginnings of tissue organization that grew more elaborate in other phyla. As either a sessile (attached) sac or a free-floating bell, they have but a single opening through which enters what is to be eaten and exits what is to be discarded. Their nervous system, a simple network, slowly transmits contraction pulses.

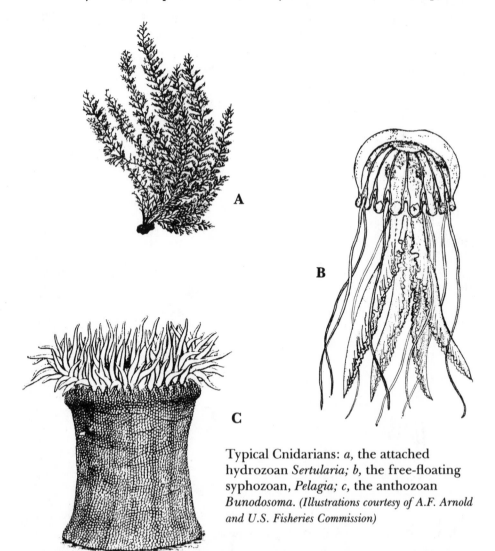

A

B

C

Typical Cnidarians: *a,* the attached hydrozoan *Sertularia; b,* the free-floating syphozoan, *Pelagia; c,* the anthozoan *Bunodosoma. (Illustrations courtesy of A.F. Arnold and U.S. Fisheries Commission)*

As if in compensation for this sluggish arrangement, they come equipped with a lightning-fast and lethal surprise package. All have stinging cells, usually concentrated along and at the tips of tentacles that surround the mouth of the sac or the rim of the bell. These fire automatically when brushed against prey, inflicting a paralytic poison and holding it fast.

The ferocity of their sting varies among species, and is geared to overcome anything from a minute copepod to a fish. To humans, most can do little more than irritate the skin, but a few can deliver a debilitating, even lethal, wallop. The nip of a Portuguese man of war is as nasty as a shock from a live lamp socket, and its aftermath is a searing burn.

The sac-shaped creatures can be soft-bodied, as are the anemones, or the tube can be surrounded by hard calcium deposits, as are the corals, who may be solitary but more often grow together in colonies.

Among the jellyfishes, class Syphozoa, the bell (*medusa*) form predominates. In the Anthozoa, anemones and corals, a sac or tube (called a *polyp*) is the dominant shape. In the Hydrozoa, either medusa or polyp can predominate or exist sequentially within a single species, each generation altering from one to the other.

Small sea anemone are good additions to a saltwater aquarium

CTENOPHORA: COMB-JELLIES

Much like the jellyfishes and, at one time, lumped with them, the comb-jellies do not have stinging cells. They capture prey with adhesive tentacles. They do not move by pulsating, but by the oarlike actions of eight rows of comb plates that beat in synchronous waves. Each comb plate is constructed of thousands of fused cilia so fine in structure that they diffract light and are beautifully iridescent in strong sunlight. Comb-jellies are bioluminescent, emitting flashes of light when disturbed.

A Bagful of Worms

Over half the phyla found in the sea have members whose bodies are worm-shaped. At one time they were all lumped in a single phylum, the Vermes, but are now separated into flatworms (Platyhelminthes), ribbon worms (Nemertea), roundworms (Nematoda), spiny-headed worms (Acanthocephala), peanut worms (Sipuncula), spoon worms (Echiura), beard worms (Pogonophora), segmented worms (Annelida), horseshoe worms (Phoronida), arrow worms (Chaetotgnatha), acorn worms (Hemichordata), and a few other lesser groupings. Some are microscopically small, others are free-living, still others parasitic. Larger free-living worms are soft-bottom burrowers, but many are good swimmers.

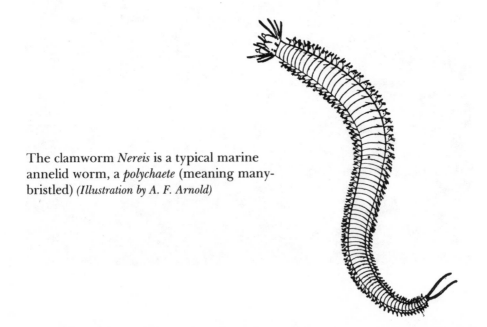

The clamworm *Nereis* is a typical marine annelid worm, a *polychaete* (meaning many-bristled) *(Illustration by A. F. Arnold)*

PLATYHELMINTHES: FLATWORMS

Simplest in design, the flatworms, like the Cnidaria, have a digestive cavity with a single opening that takes in food and expels waste. Unlike Cnidaria, their excretory, muscular, nervous, and reproductive systems are arranged as discrete organs within a substantial frame.

Of the three classes of flatworms—Trematoda (flukes), Cestoda (tapeworms), and Turbellaria (free-living flatworms)—only the latter have members living openly that are large enough to see underwater. Even so, polyclads and triclads are *cryptic,* living under stones or well camouflaged on the surfaces of larger invertebrates. They can swim well and a few live in open ocean.

NEMERTEA: RIBBON OR PROBOSCIS WORMS

Ribbon or proboscis worms are long, soft, unsegmented, and often vividly colored, found in mud, under stones, among seaweeds, and floating in open seas. They are the next step up from the flatworms in biological complexity. Their digestive system has two openings, allowing a one-way flow for food. They also have an elementary circulatory system, although it lacks a pump.

To catch food they can extend a long probe, barbed and poisonous in some species, which they quickly wrap around their prey, exuding a toxic mucus. About nine hundred species have been identified worldwide.

NEMATODA: ROUNDWORMS

Important because they cause so much disease among animals, plants, and humans (hookworm, Ascarisis, and filarial afflictions), only a few marine roundworms can infect man. Some parasitize fish, sea mammals, and invertebrates, while others live free in bottom sediments. All are small, and many are microscopic.

ANNELIDA: SEGMENTED WORMS

The body of an *annelid* is subdivided into like, but not identical, repeating units. The organs are well defined: a series of hearts circulate the blood; the digestive tract is differentiated into a pharynx, crop, gizzard, and intestine; and the brain controls a central nervous system.

Their remarkable contribution to evolutionary development and an essential step to the higher invertebrates is the complete separation of their digestive tract from their body wall, creating a fluid-filled space in between, the *coelom*. Thus the outside of the animal is unlinked from what goes on inside, allowing vast new possibilities in form and function.

Most of the ten thousand or so species of annelids are not the familiar garden variety earthworms nor the leeches, but the bristled marine polychaetes. Some are swimmers or crawlers: the clamworm, for example. Others dwell in tubes and include such pretty examples as the feather dusters and Christmas tree worms of the tropics.

SIPUNCULA AND ECHIURA: PEANUT AND SPOON WORMS

These phyla were once classified together and are thought to be closely related to the Annelida but are unsegmented. They are burrowers who keep their body under the sediments while extending a feeding and breathing tube into the waters overhead. All are suspension or deposit feeders. Worldwide, 300 species of peanut worms and 140 species of spoon worms are known.

CHAETOGNATHA: ARROW WORMS

Exclusively marine, arrow worms are good swimmers, more fishlike than wormlike. Equipped with strong, bristled jaws, they prey upon the earliest stages of fish and, in turn, are the prey of larger fish. They often turn up in summer and fall plankton tows.

HEMICHORDATA: ACORN WORMS

Acorn worms burrow in sandy or muddy bottoms. The adult body has three distinct parts: a carrot-shaped proboscis attached to a ringed collar on an elongate, somewhat flattened body.

It is their larvae that intrigues the systematist: They seem to be a link between the chordates (in which humans find themselves) and the echinoderms (in which the starfishes are found).

Animals: Higher Invertebrates

With the new freedom provided by the coelomic cavity, evolutionary design blossomed in the mollusks, the arthropods, the echinoderms, the chordates, and a few other odd phyla. New systems emerged; some persisted on, others waned. The marvels we see in today's vertebrates—keen senses, muscular agility, and complex behavior—had their antecedents in these forms.

MOLLUSCA

The range of progress in this phylum is extraordinary, from the primitive and sluggish chiton to the quick and bright octopus. The generalized body plan includes a head; a broad muscular foot; a visceral mass containing well-developed organs of circulation, digestion, excretion, and reproduction; and a mantle that, in many species, secretes a shell. However, broad generalizations beget exceptions. You will be hard-pressed to find the head in a bivalve or the foot in a cephalopod. In the former, it's virtually nonexistent, and in the latter, it has modified into eight arms.

Second only to the insects in number of species, the hundred thousand or so members of this phylum fall into seven classes, of which four are found in shallow marine waters: Polyplacophora, the chitons; Pelecypoda, the bivalves; Gastropoda, the snails; and Cephalopoda, the squids, cuttlefishes, and octopuses.

Chitons are the most primitive of the living mollusks. They have a flattened oval body covered either by a tough cuticle or eight jointed plates. A foot covers the underside and allows them to cling to rocks in pounding surf. Most feed by scraping algae, but some capture living prey.

The *bivalves* are two-shelled. They feed by filtering food particles from water passed over their gills. Their foot can probe, burrow, glide, anchor, or bore, depending on the species. Some can cement or tie themselves down; others can swim by rapid "clapping."

They include nut clams, ark shells, scallops, mussels, jingle shells, oysters, pens, wing shells, file shells, spoon shells, duckbills, pandoras, black and brown clams, cockles, venus clams, rock borers, tellins, razors, surf clams, soft-shelled clams, angel wings, piddocks, and shipworms.

The body plan of the *gastropods* contains a curious twist called *torsion:* the viscera has undergone a 180-degree turn, placing most of the body over the head (which means waste exits very close to the gills!). They have two stalked eyes on their head and a *radula,* a rasplike device, for scraping and boring. The foot is well suited for clinging and creeping. They make up nearly eighty percent of living mollusks—about half are marine.

The gastropods include limpets, top shells, wendletraps, moon shells, slippers, sundials, periwinkles, worm shells, conchs, helmets, tuns, cowries, sea snails, dove shells, mud snails, augers, dog whelks, olives, and sea butterflies. A few come without a hard housing: sea hares and the nudibranchs, or sea slugs.

The surprise in the Mollusca are the *cephalopods:* squid, cuttlefish, sea arrows, octopuses, and argonauts. They are remarkably well coordinated, with keen senses and quick reflexes. Few creatures are more enjoyable to watch underwater than the octopus. In the daytime these wily and elusive creatures will watch your every move, backing off and darting away at the slightest sign of threat, but at night, in the beam of a diver's light, they appear totally bewildered and will allow you to pick them up gently.

ARTHROPODA

The joint-legged animals contain most of the world's known species. An enormous phylum with diverse members, its numbers are dominated by the insects, of which only a few are marine and of little interest here.

All *arthropods* have a jointed outer skeleton made of layers of chitin (a horny substance chemically similar to fingernails), protein, wax, fat, and in some of the more heavily armored members, calcium carbonate. Their bodies are segmented like the annelids from which they arose but, unlike the annelids, each segment is distinctly different, separated into a head, a thorax, and an abdomen.

Their paired limbs provide them with a multiplicity of functions: antennae for sensing, mouth parts for biting and grinding, pincers for grabbing and feeding, and legs for breathing, walking, and paddling.

Opposing pairs of muscles articulate these armored limbs: flexors and extensors provide movement in a way quite similar to the way our own muscles move our arms and legs.

The subphylum Crustacea, about thirty thousand species, contains four important marine classes: the Ostracoda, Copepoda, Cirripeda, and Malacostraca.

Ostracods are small (less than one-tenth of an inch), their bodies protected by a bivalve shell. They live among seaweeds, in sponges, and in coarse sediments.

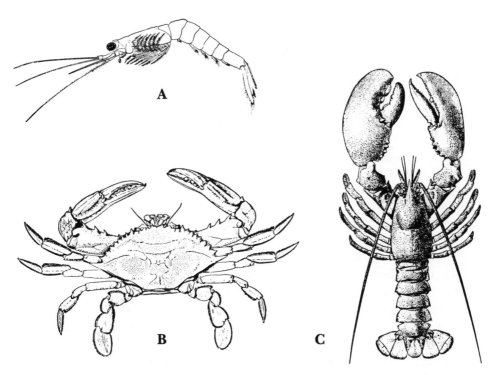

Familiar Crustaceans: *a,* the mysid shrimp, *Mysis sternolepsis; b,* the blue crab, *Callinectes sapidus; c,* the American lobster, *Homarus americanus. (Illustrations from A. F. Arnold)*

Copepods are planktonic free-swimmers. They are the single most important source of food for fish in the sea. The number of individual copepods in the sea probably exceeds that of all other multi-celled animals on earth.

Cirripedia, the acorn and goose barnacles, encased in shells of calcium carbonate, live attached to fixed and floating objects. They strain nutrients from the water by extending and retracting curved, feathery appendages that open and close like grasping hands.

Malacostraca include the mysids, who are pelagic swimmers; the isopods, mostly free-living but some parasitic (fish lice) and a few wood borers (gribbles); amphipods, represented by sand hoppers, beach fleas, skeleton shrimp, and whale lice; and the decapods, which include true shrimps, ghost shrimps, hermit crabs, true crabs, lobster, and others.

Another subphylum, the Chelicerata, contain the horseshoe crabs, sea-spiders, and mites.

ECHINODERMATA

Entirely marine, the phylum includes sea stars, brittle stars, sea urchins and sand dollars, sea cucumbers, and the rarely seen sea lilies. A zoologically puzzling group unlike other animals, they share some of the attributes of the chordates, but the true meaning of these relationships has been the topic for lively disputes.

Echinoderm means "hedgehog skin," referring to the tough, spiny outer covering that is sometimes reinforced with imbedded calcite crystals. The five-sided radial symmetry has produced a body that bears no clear-cut head; the mouth may face up, down, or sideways.

All echinoderms have a hydraulic vascular system that serves for respiration as well as movement. Taking water in through a sieve plate, pumped by beating cilia through a system of canals, it's used to extend or withdraw tube feet. Each tube foot can act as a suction cup, collectively exerting enormous force. The tube foot also serves as a sensory organ.

The phylum is divided into five classes: the Asteroidea, sea stars; Ophiuroidea, brittle stars (some zoologists lump sea stars and brittle stars together, calling the combined class Stelloidea); Echinoidea, sea urchins and alike; Holothuroidea, sea cucumbers; and Crinoidea, sea lilies.

Sea stars have five or more arms that radiate from a central disk that contains a mouth on the same side as the tube feet and a sieve plate on the opposite side. The skin on the upper side may also be outfitted with sets of small grasping pincers that keep its surface free of settlers.

An eyespot sits at the end of each arm. Sea stars move more slowly than brittle stars, in part because their arms are heavier, shorter, and meet together at the central disk. Brittle stars have long, flexible, un-branched arms. Basket stars have branched arms.

Unlike sea stars, the sea urchins, heart urchins, cake urchins, and sand dollars do not have arms. Their body is encased within a *test,* a perforated skeleton made of unmovable radial plates fitted with ball joints that hold spines provided with sockets. The plate perforations serve a series of functions: excretion, water intake, and outlets for tube feet.

The sea cucumbers are cylindrically shaped, generally leathery creatures with a mouth at one end and an anus at the other. Feeding tentacles (enlarged tube feet) surround the mouth, taking in sus-pended particles or gulping soft sediments as the animal burrows or

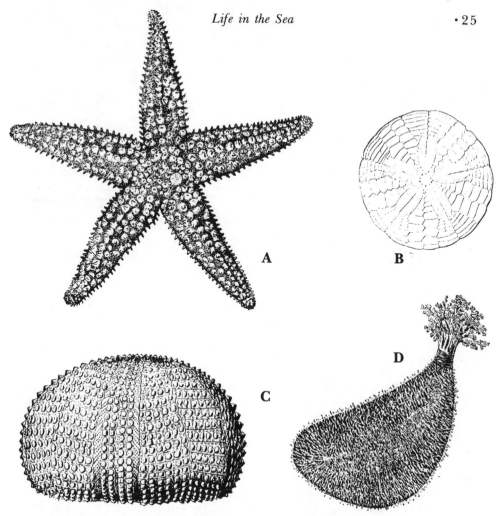

Typical Echinoderms: The five-rayed symmetry of the echinoderms is obvious in *a,* the sea star *Asterias forbesi;* and *b,* the dried test of the sand dollar *Echinarachnius parma;* less so in *c,* the test of a sea urchin; and not at all in *d,* the sea cucumber, *Thyone briareus.*

makes its way over loose surfaces. Rows of tube feet also provide locomotion. They have few predators. Their flesh is filled with hard ossicles of calcium carbonate and is toxic. When attacked or violently disturbed, the sea cucumber responds by everting its stomach and discharging a load of mucus, disgusting to us, but an effective means of defense.

Sea lilies and feather stars live in very deep water, but a few species are occasionally found above the hundred-foot depth. They are living fossils, remnants of the crinoids that filled the Paleozoic seas.

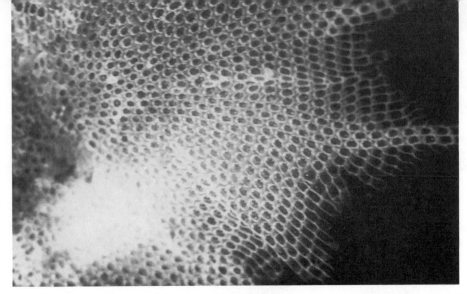

Membranipora pilosa is a bryozoan that encrusts stones, shell, and seaweed fronds with a beautiful lacelike structure.

ODD FELLOWS

The moss animals, *Bryozoa,* are sessile, mainly colonial forms that grow on rocks, shells, algae, and other objects. Their lacelike or tubed calcium carbonate houses enclose tentacles that can extend or withdraw in much the same manner as the hydroids. Some sit on individual stalks. Identifying them other than by the pattern of the structures they build requires the use of a microscope.

Just as the bryozoans can be confused with the hydroid cnidarians, so can the *Brachiopoda,* the lamp shells, be confused with mollusks. Enclosed in small, bivalved, dissimilar shells, the animal attaches to rock or into a burrow with a stalk and feeds with a pair of ciliated tentaclelike arms. Sparse in shallow water, they are more plentiful in the deeps.

CHORDATA

The *chordates* are comprised of two subphyla: the *urochordates* (containing the tunicates), and the *vertebrates.* The *tunicates,* the sea squirts and kin, are vase-shaped sessile animals with two orifices. One draws water in, the other ejects it. Their ties to the vertebrates is evident only in their larvae, not in the adult. The larvae possess a primitive *notochord,* the precursor to the vertebrates' backbone.

The Vertebrata are animals with backbones, an encased brain, and a central nervous chord running from the brain through the backbone to organs and muscles of the body.

The sea hosts many mammals, but it is the fishes who dominate the environment. Fishes are generally categorized into two groups: those with a skeleton of cartilage, and those with a skeleton of bone. The cartilaginous fishes contain about 200 species of shark, 350 species of skates and rays, and 25 species of chimaeras. Thirty times as many bony fishes exist. Most of them are ray-finned and most of the ray-fins are *teleosts*. Teleosts are modern ray-fins who, during evolution, lost the heavy armor plate that currently exists in only a handful of living species—for example, the sturgeons.

THE OVERLOOKED

Although the main animal phyla have been touched on, there remain others who, either because they are obscure or very small, have not been mentioned. Some live between grains of sand, others on or within other plants and animals, still others float free. Their kind, their ways of living, and their relationships are there to be explored by the curious and are especially suited for those who might enjoy the challenge of studying the micro world they inhabit.

The archannelid *Nerilla* lives between grains of sand. *(Photo by J. Simon)*

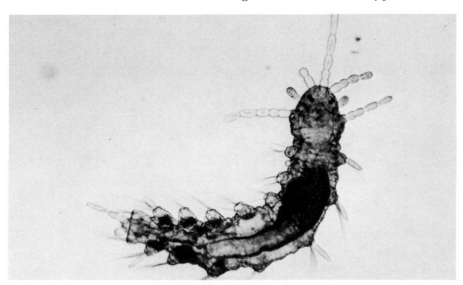

A WORD ABOUT NAMES AND MEASURES

In the following chapters, whenever an animal or plant is mentioned for the first time, both its most appropriate common name and its scientific name will be given. Occasionally, only a scientific name will be given, simply because there is no common name.

Up to this point, all temperatures, lengths, areas, and weights have been given in system units used in the United States. Other than Libya and Algeria, we are the only nation still using this system. The metric system has replaced it and become standard almost everywhere else in the world, including American scientific literature.

In subsequent chapters, units will be given in the metric system, and often, though not always, followed by the U.S. equivalent. You cannot expect to have the same familiarity with these units as you do with inches, feet, and pounds, but, with a little practice, you can become accustomed enough so they relay a meaningful message to you.

Look over the brief list of conversions given in the Chapter Notes at the end of the book. Try to remember the approximate equivalents—that's close enough for most purposes. You will still get the point if someone remarks that "A gram of prevention is worth a kilo of cure."

2

At Sea: Pelagic Swimmers

A T SEA, FROM THE DECK OF A SHIP, THE SIGHT OF FINNED swimmers alongside is heralded as a special event. Flying fish skittering away from the ship's track or gulls screaming over baitfish driven to the surface by undersea hunters all warrant excited attention and comment. Fishing feeds the imagination as well as the body. A school tuna hits a plug trolled at ten knots. You marvel at the burst of speed it took to snatch that lure and the strength the fish will exert in its fight to live. Or you slowly troll a heavy line in a chum slick and hook a shark. Hauled alongside, he is tagged and released, and you ponder how far he will travel and his ultimate fate.

If you make trips enough and get into the water frequently enough, you will see some of the pelagic (ocean-dwelling) wanderers who regularly come inshore following the courses of the great currents and abundant food. Some are on their way to traditional breeding areas, others to seasonal feeding grounds. You may only catch a glimpse of them. But that quick look at even a small fraction of their existence will heighten your understanding of the cycles of life in the sea and raise your appreciation of the difficulties in unraveling the life histories of these creatures.

THE TEEMING BILLIONS

These are the middle link in the oceanic food chain, the great schools of pelagic wanderers—herrings, anchovies, mackerel, and the like—who feed upon the drifting plankton of the sea and, in turn, are the prey of the faster swimmers and man.

And man's share looms larger every year. The annual harvest of Atlantic menhaden, herring, mackerel, and Pacific jack mackerel collectively exceeds all other fisheries combined. They end up canned, as fish meal, oil, animal-feed supplement, and pet food. This supply is not endless. The vagaries of nature coupled with overfishing can cut some stocks below the numbers essential for annual renewal, reducing the total number drastically. What will be the ecological consequences of this to the balance of life in the sea?

Twenty-seven species of herring and sixteen species of anchovy swim in our waters. The herrings include shad, alewife, menhaden, pilchards, and sardines, as well as true herrings. All have one dorsal fin, soft-rayed fins, deeply forked tails, either no teeth or very small ones, deep flattened bodies, and scales that come off at a touch.

Their life histories differ substantially. The **American shad,** *Alosa sapidissima,* is anadromous; that is, it enters freshwater rivers from the sea and swims upstream to breed. The run starts in the spring when the water temperature reaches 12° C (53° F). Atlantic Coast shad runs begin in Georgia rivers in January, then progress northward from Delaware to Maine in May and June. The shad spawn in sandy or cobbly shallows. Once done breeding, the spent fish return to the sea.

Most herrings spawn at sea. **Atlantic menhaden,** *Brevoortia tyrannus,* also known as pogy, mossbunker, fat back, and some thirty other common names along the Atlantic Seaboard, travel in enormous schools feeding on plankton. They spawn at sea during summer in the Northeast. Their eggs float. The fry grow up to six to eight centimeters in length by their first winter. Adult numbers fluctuate tremendously, in part from overfishing and in part from natural predation.

The true herrings, the *Clupea* species, spawn in shoal waters, laying their eggs over gravel or rocky bottom. The eggs sink and stick to rock, weed, sand, and one another. The waters of Maine, from five to one hundred meters (about fifteen to three hundred feet) deep, are the chief spawning grounds of the **Atlantic herring,** *Clupea harengus harengus.* On the Pacific Coast, the **Pacific herring,** *Clupea harengus*

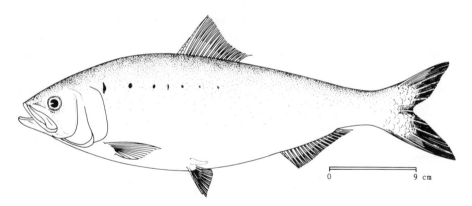

American shad, *Alosa sapidissima (Illustration courtesy of Fish and Agriculture Organization—The United Nations)*

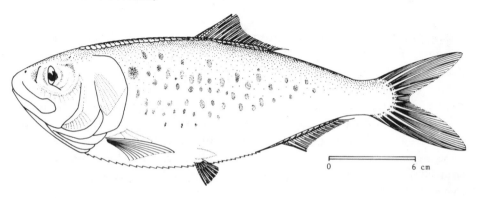

Atlantic menhaden, *Brevoortia tyrannus (Illustration courtesy of Fish and Agriculture Organization—the United Nations)*

pallasi, breeds in bays during winter and spring (the two herrings, once thought to be distinct species, are now considered separate races). To the south, the **Pacific sardine,** *Sardinops sagax,* spawns in spring and early summer off Baja and Southern California.

The success of a spawn depends on the volume of enemies present when the fish are still in the larval and fry stages. Predatory comb-jellies and arrow worms can decimate fry numbers if they are simultaneously abundant.

Overpredation of the young or overfishing the adults can cause catastrophic declines in sardine numbers. So can a change in water conditions. The Pacific sardine fishery, so prolific during the 1930s and 1940s, collapsed in 1947 primarily because of a slight shift in water temperature.

Along our Northeast Coast, **Atlantic mackerel,** *Scomber scombrus,* announce the arrival of spring. Moving inshore and northward as water temperatures rise above 8° C (46° F), the vast schools feed and spawn on the move. By June they have reached the waters off Cape Cod. Spawning goes on well into summer. A mature female will lay half a million eggs a season for four seasons. To break even and maintain the species, one egg in a million must reach maturity. Every few years one year-class (the survivors of a specific year's spawn) will do exceptionally well, beating the odds and swelling their ranks enormously.

Mackerel are high-speed swimmers. They are heavier than water, and thus must swim or sink. Buoyancy in some fish is controlled by a gas bladder, which is missing in mackerel. This bladder, an evolutionary remnant of the lung, is a mixed blessing. By slowly adjusting the amount of gas in the bladder, the fish can maintain a state of weightlessness, much like a submarine with ballast trimmed. Unlike the sub, though, if the fish rises too quickly, the gas expands more rapidly than the fish can reabsorb it. Bloated fish brought up by trawl nets suffer this fate. The point is that fish with gas bladders can't change depth quickly, but mackerel are not hampered by this restraint. A school swimming near the surface can vanish into the depths within minutes, to reappear on the surface miles away and hours later.

Squid swim the oceans and coastal seas in vast numbers, feeding on small fish and invertebrates, and are themselves a mainstay for larger, faster fish and mammals.

They are cephalopods, near kin to the octopus, cuttlefish, and nautilus, and are the most advanced of the mollusks. Their sense of sight is as good as that of a fish, and they are every bit as agile. Capable of instant action, their two longest tentacled arms can seize a moving target with startling alacrity, and their jet propulsion can send them darting instantly away in pulsed flight.

Their siphon can reverse direction, propelling them forward or backward. With two finlike flaps and the use of their siphon, they can hover or move slowly or rapidly in either direction. Small groups usually swim in parallel rows.

The common squids of American waters are the **longfin squid,** *Loligo pealii;* the **northern shortfin squid,** *Illex illecebrosus,* on the East Coast; and the **market squid,** *L. opalescens,* of western waters. They breed in groups. Females lay cigar-sized capsules containing fifty or more eggs, attaching them to anything convenient, creating a cluster

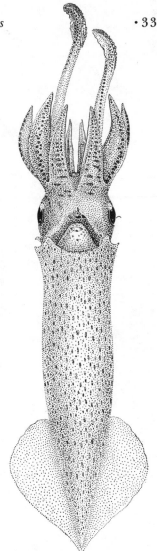

Market squid, *Loligo opalescens (Illustration courtesy of
U.S. Bureau of Fisheries)*

of a hundred or so capsules that look like the end of an old-fashioned
dust mop. The eggs are laid in winter and take thirty to forty days to
hatch. If you find them, you can keep them in an aquarium and
watch them hatch. Hatchlings move by flexing the *mantle*: a cloak of
flesh covering the body. Their eyes are fully developed at birth; they
can see food and avoid enemies. Their pigment cells are not under
full control upon hatching; the cells expand and contract but the
overall result doesn't produce a color change as it does in adults.

Large squid with body lengths of 1.2 meters (4 feet), which means
an overall length in excess of 3 meters, occasionally get caught in nets

Mass spawning of Pacific squid. The gelatinous bundles of egg cases are attached to one another. *(Photo by E. S. Hobson)*

or hooked by fishermen. Now and then the remains of a giant deep-water squid washes up or is snagged. These creatures are the largest living invertebrates. One has been found with a body length of seven meters, suggesting an overall length of seventeen meters (fifty-five feet) and a weight of thirty tons. Pieces of others suggest that still larger specimens exist.

Schooling

Few experiences rival swimming within a large school of small, fast-moving fish. Comprehension of numbers is sorely tested when confronted by these multitudes in swift motion.

Herrings, sardines, and their relatives school in vast numbers. Schooling (individuals acting in "egalitarian synchrony," as one authority has put it) is a widespread phenomenon among fish: Half of all juveniles and a quarter of all adult species school.

A school of fish is more than a milling throng in close proximity. All individuals face the same direction, swim at the same speed, space

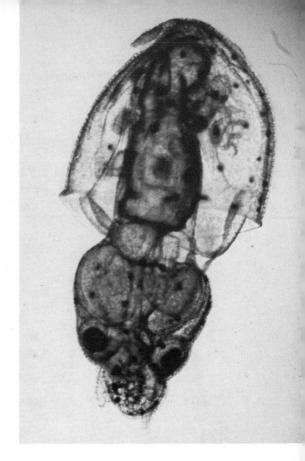

A newly hatched squid just one-quarter inch long. The spots are *chromatophores* that will allow the adult to change color.

themselves regularly between neighbors, and act in unison. Each member of the school is spatially aware of its immediate companions and instantly adjusts to their movements. Young fish space themselves one-half to one body length apart, parallel to one another, and midway between the two parallel fish leading them. The hydrodynamics of this arrangement allows each fish to avoid the drag eddies created by the leading fish.

They are guided by sight and lateral-line sense. The *lateral line,* a series of pressure-sensitive cells around the head and down each side of the body, keeps schooling fish aware of spacing and sudden turns. Sight plays a important role. Fish tend to disperse in total darkness but will school in moonlight.

What good does schooling do? One theory has it that it helps maintain cruising speed. Skin mucus slowly dissolving away from each fish acts as a drag reducer in much the same way that adding minute amounts of certain surface-active agents to water allows the water to

Fish school in shallows as well as at sea. These dwarf herring, *Jenkinsia*, are common in tropical shoal waters.

slip through pipes and hoses faster. Also, schools contain males and females of the same year-class, therefore the chances of successful mating, either by pairing or mass spawning, increases significantly.

Does schooling help thwart enemies? Opinions are mixed. Given limited visibility, one theory goes, the probability of a single fish being detected is less in a group than as a scattered individual. Another theory postulates that what counts is the confusing appearance of the school. Hunters have only so much energy to make their moves and must home in on one individual to increase the odds of success. Any visual diversion that deflects the charge promotes a miss.

Disorientation upon first swimming within a large, enveloping bait school will do more to enlighten the naturalist about visual bafflement than a thousand words. One is surrounded by an envelope of empty water walled with streaking corridors of fish, that, acting synchronously, flash a confusing pattern of light and dark in near perfect unison. They all seem to be swimming forward with you, but soon a hole of clear water opens and enlarges ahead. Suddenly they are gone, either behind you or headed off in another direction.

What happened? As you swim forward, the leading fishes, well ahead of you and out of your sight, turn back along the outside of the advancing school. Taking more and more of the outsiders with them, they re-form up at the rear of the school. This process eventually peels away all those ahead of you and regroups them behind you.

FAST SWIMMERS

One of the truly memorable underwater sights is a glimpse of a fast loner: a large oceanic wanderer who feeds on small fish and squid in open midwater and is built for speed and endurance. Water is eight hundred times denser than air and muscling through it quickly requires power and a streamlined body shape.

Fast fishes have a *fusiform* (tapering toward each end) shape and a slim tail that holds a large crescent-shaped fin. This design has had a certain hydraulic universality over eons, from the ancient ichthyosaur to the latter-day mackerel shark and tuna. Broad bands of muscles flex the body from amidships to tail and supply the motive force. Tuna can cruise at ten to fifteen knots and have been clocked at over forty knots in short bursts. (A knot is one nautical mile—1.15 statute miles—per hour).

Top speeds of fast fishes range from nine to thirteen body lengths per second up to an overall body length of two meters, but this can be sustained for no longer than a few minutes. Absolute speed is limited by body length; thus, a twenty-centimeter long herring going flat out rarely exceeds five knots, while a two-meter tuna can exceed forty knots.

As the fish pushes into new space ahead of it, turbulent water fills the vacated space, creating resistance to its forward motion. A streamlined shape minimizes this effect but does not eliminate it. But the influence of drag decreases on a longer body. Fish can reduce drag more efficiently than a human-designed ship can because a fish's skin is flexible where the ship's hull is rigid. A pliable skin lets the water slip over a longer portion of the body before it breaks away, to form eddy currents that hinder forward motion.

Fast fish swim with short, stiff, powerful side-to-side strokes. Swimming style and lateral body flexibility go hand in hand; the more flexible the body, the more exaggerated the S-shaped undulations and the more inefficient the effort.

What of the fins? They provide stability, not power. They compensate for the yaw, pitch, and roll imposed by swimming and water movement. A sea bass, using its large pectoral fins, can hover effortlessly by sculling forward or backward. Pectoral fins also provide steering. To bank into a turn, a shark lowers the pectoral fin on the side to which it intends to go. Fast fish often retract their pectoral fins as they pick up speed. The skipjack can tuck its fins in very close to its sides, rotating them as they retract, increasing their attack angle to the water, and reducing drag.

Tuna, swordfish, and billfishes are the fastest swimmers along our coasts. Among the big tunas that visit both the Atlantic and Pacific coasts are the **skipjack,** *Euthynnus pelamis;* the **yellowfin,** *Thunnus albacares;* the **albacore,** *T. alalunga;* and the **bigeye,** *T. obesus,* which is the largest.

Both oceans have their own bonito: the **Pacific bonito,** *Sarda chiliensis,* and the **Atlantic bonito,** *S. sarda.* Atlantic waters are also home to the **striped bonito,** *S. orientalis,* and the **wahoo,** *Acanthocybium solanderi,* which also plies Pacific waters off Mexico.

One species of **swordfish,** *Xiphias gladius,* swims both coasts, as do two worldwide representatives of the billfishes, the **sailfish,** *Istiophorus platypterus,* and the **blue marlin,** *Makaiara nigricans.* The **white marlin,** *Tetrapturus albidus,* only visits our Atlantic Coast, while the **striped marlin,** *T. audax,* confines itself to the Pacific.

The scientific name of a fish is especially valuable to know if you are dealing with oceanic species. Common names of oceanic gamefish vary so widely that it's difficult to tell just what fish is meant. The bluefin tuna is a *tunny* to an Englishman and a *horse mackerel* to many others.

The tunas are among the most streamlined fish in the sea, and wander worldwide. Identifying a member of the tuna family is easy. Its sleek body shape, crescent-moon tail, and *scutes,* or finlets, near the tail fin are unmistakable signatures. Telling one species from another in the water is another matter, however. The *dorsal* (top) and *ventral* (forward bottom) fins, so prominently displayed in book illustrations and fish seen dead at the dock, fold back into grooves as the fish swim. One is left to look for barred markings, the relative length of the pectoral fins, and other more subtle clues that are not at all easy to glean when the fish is at high speed.

Chances of seeing a tuna are best around schools of their prey. Get close to a school of herring, bluefish, or mackerel. The tuna following

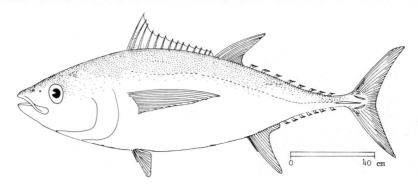

Bigeye tuna, *Thunnus obesus*

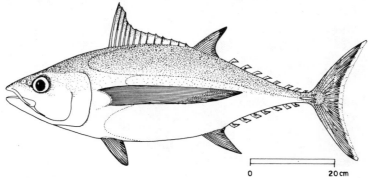

Albacore, *Thunnus alalunga*

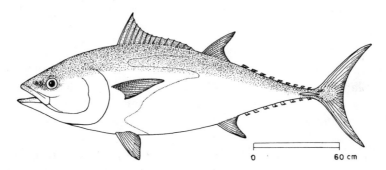

Northern bluefin tuna, *Thunnus thynnus*

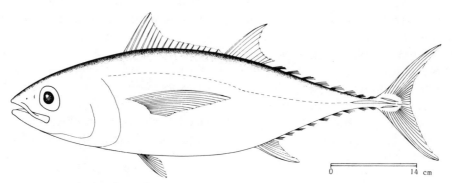

Blackfin tuna, *Thunnus atlanticus*

(Illustratioons courtesy of Fish and Agriculture Organization—the United Nations)

A white marlin is tagged by a sport fisherman in a cooperative program with the National Marine Fisheries Service. *(Photo courtesy of National Marine Fisheries Service)*

them will often pass by to look you over. If you want photographs of them, plan for split-second shooting. You will have less than ten seconds to get off a shot, given eight-meter water visibility. A pod of two or three can appear out of nowhere in an instant, then vanish, leaving you dumbstruck.

Tuna cover staggering distances in a relatively short time. Six large tuna tagged in the Straits of Florida wound up in Norwegian waters; four were recaptured 50 to 119 days after release, having logged forty-two hundred nautical miles. Another giant, also tagged in the Florida Straits, went south, crossing the equator. Young **bluefin tuna,** *Thunnus thynnus,* tagged between New Jersey and Cape Cod, migrated to the southeast corner of the Bay of Biscay within weeks. This and other evidence suggests the young tunas' oceanic paths are distinctly different from older, larger fish. Couple that complexity with evidence for different migration patterns for separate stocks, and you can understand why the fisheries biologist often needs the help of the sport fisherman.

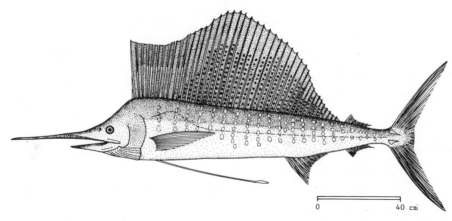

Atlantic sailfish, *Istiophorus platypterus*

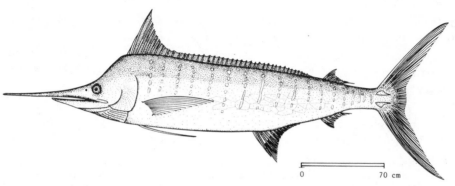

Blue marlin, *Makaira nigricans*

(Illustrations courtesy of Fish and Agriculture Organization—the United Nations)

FISH TAGGING

Big-game sport fishermen have lately turned to tagging and releasing their catches. Most of the long-distance migration information about tuna, swordfish, and marlin has come from cooperative tagging programs involving sport fishermen.

Volunteer tagging programs grow more important every year as federally funded programs dry up, but the information becomes vital to decisions on stock management.

The largest private tagging program in the United States is run by the American Littoral Society at Sandy Hook, New Jersey. Since the program began in 1965, the Society has distributed over 230,000 tags. About half have been put on fish.

Returns run about eight percent. Over eighty species have been tagged, predominantly striped bass. The return data is computerized and used by a number of institutions. The Society distributes the tagging kits for a nominal fee, which includes tagging instructions, measurements needed, and how to handle the fish to maximize its survival. (See the Chapter Notes for information on how to contact the Society.)

For tuna, marlin, swordfish, and other big oceanic species, the National Marine Fisheries Service southern fisheries centers operate a cooperative gamefish tagging program on both seaboards. In 1982, thirty-six hundred fish were tagged and released, and in that year ninety-five were recovered.

The centers will supply the tagging equipment and advice on the proper way to tag. They also circulate a newsletter that occasionally contains requests for unusual specimens needed for age and growth research. (The addresses are given in the Chapter Notes.)

For shark, the tags used on smaller, soft-bodied fish and the accompanying methods of tag implanting are unsuitable. For starters, the shark must be tagged in the water with the cooperation of two or three people who know what they are doing. The National Marine Fisheries Service at Narragansett, Rhode Island, runs a cooperative shark-tagging program and will supply tags, advice, and a shark newsletter.

If you want to tag shark, learn about the shark species in your area and go fishing with someone who can identify shark in the water and who knows how to fight them without drowning them. If all this seems too complicated, consider that ninety-five percent of all the shark in this program were caught and released by volunteer taggers. (See the Chapter Notes for information on how to contact the Service.)

SHARKS

Sharks, the dark spectres of the ocean, are universally held in ill repute. The ferocity with which they slash and bite out a hunk of prey with a deft snap of their jaws creates an image instantly remembered when an ominous gray form glides gracefully out of the underwater gloom. The reality that humans as a group suffer little damage from

Tag No.

....

Date Caught ....
Species of Fish ....
Location Released ....
Approx. Length .... Weight ....
Taggers Name ....
Address ....

....
Phone (   )...

Push metal needle with tag
through this area of fish.
Remove needle, tie overhand
knot in tag and release fish.
↓ SAVE NEEDLE

Special Comments

....
....
....

TRIM EXCESS TAG
....
Signature

FISH TAGGING

Tagging a fish is simple with an American Littoral Society tagging kit. The tag is a length of thin, bright yellow flexible plastic tubing with a return name, address, and identification number imprinted on it. The tag slips inside a hollow stainless steel needle. Needle and tag together are pushed through the body of the fish about an inch below the top of its back, just aft of its last dorsal fin. The needle is slipped off the tag, leaving both ends of the tag free on each side. Tie the free ends together with an overhand knot, leaving as much space as possible to allow for growth. Measure the length of the fish from its snout to the fork of its tail. An accompanying card, which is mailed back to the Society, asks for information on species, and where caught, as well as length.

Successful taggers have come up with a list of suggestions that help insure the survival of the fish:

- The fish must be at least eight inches long.

- Use single hooks, not trebles. File down barbs. Keep the hooks and the tagging needle sharp by frequent honing.

- Use big tackle that can haul the fish in before it is totally exhausted.

- Lay the fish on a wetted surface and cover its head with a wet cloth while dehooking, measuring, and tagging it.

- Get the hook out as gently as possible. Don't tag a badly gaffed fish or one that has swallowed the hook.

- After tagging, pick it up by its undersides and work it through the water until it is ready to swim away.

- With big bluefish, hold them by the tail and take out hook with a hook remover. With amberjack, gaff the lower jaw and gently lift it to the deck.

- Don't use a net to board the fish. The fish suffers too much stress untangling it.

sharks offers stark comfort at such a moment—we are conditioned to the myth of their menace. It is true enough that sharks can threaten us, but that threat has been exaggerated and can be met by good sense and preparation.

The elasmobranchs—sharks, skates, and rays—are cartilaginous fishes: their skeletal structures contains no hard bone except for teeth and small toothlike scales in their skin. The largest, the **whale shark,** *Rhincodon typus,* can grow to thirteen meters in length. The **basking shark,** *Cetorhinus maximus,* is also huge, seven to ten meters long. Both are harmless slow swimmers, and can be safely approached in the water.

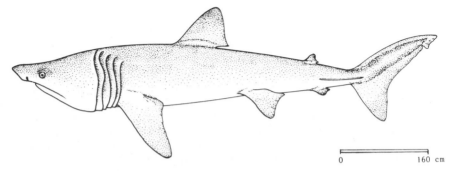

0   160 cm

Basking shark, *Cetorhinus maximus*
(Illustration courtesy of Fish and Agriculture Organization—the United Nations)

But make sure it's a whale shark or basking shark before you enter the water. The **white shark,** *Carcharodon carcharias,* has a similar body shape and *is* dangerous. The skin of the basking shark is dark gray to brown and is covered with light irregular patches, while the great white is evenly shaded. Compared to the white, the five gill slits of the basking shark are huge, extending vertically down both sides of the body. Both the whale and basking sharks are filter feeders and are equipped with enormous gill rakers, bristlelike sieves that line their gill arches.

Sharks have much larger fins than the fast bony fish, but those fins are more efficiently streamlined. The fins are always erect. The pectoral fins can be pivoted, dropped, or raised for turns or changes in depth but cannot be used for hovering. A shark's smooth, gliding, effortless motion under the water is a visual delight, the quintessential expression of the right design for the open sea. The **blue shark,** *Prionace glauca,* has a body design and propulsive ability that has been compared to a modern submarine and judged six times more effi-

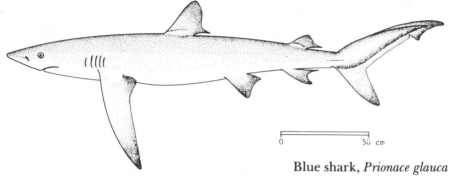

Blue shark, *Prionace glauca*
(Illustration courtesy of Fish and Agriculture Organization)

cient. A two-meter, thirty-kilogram blue can cruise at twenty knots and spurt to forty in short bursts.

The shape of the tail fin of a shark differs from that of a bony fish. The forks, or lobes, of bony-fish tails are almost always of equal length and area, while the upper lobe of the tail of a shark is always longer than the lower lobe. This design difference is carried to extremes in thresher sharks, whose sicklelike upper tail lobe is as long as their body length. They make good use of it for feeding. A group of them will circle a school of small fish swimming near the surface and leap from the water while thrashing their tails. This frightens the school, which tightens into a more compact mass. One by one, the threshers charge, mouths open.

The unevenly lobed tail might lead you to believe the shark is a slow swimmer, but stomach contents of the **oceanic whitetip shark,** *Carcharhinus longimanus,* have revealed squid, barracuda, tuna, white marlin, and other fast fishes. The oceanic whitetip is pelagic, found only in waters over two hundred meters deep and warmer than 22° C (72° F). Whitetips do considerable damage to longline catches of tuna and swordfish and are a danger to divers. They are one of the commonest sharks in open tropic seas.

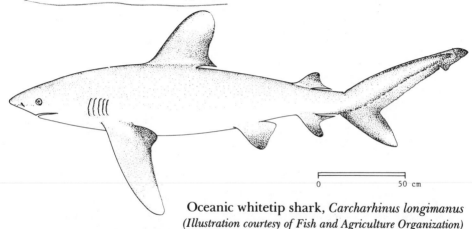

Oceanic whitetip shark, *Carcharhinus longimanus*
(Illustration courtesy of Fish and Agriculture Organization)

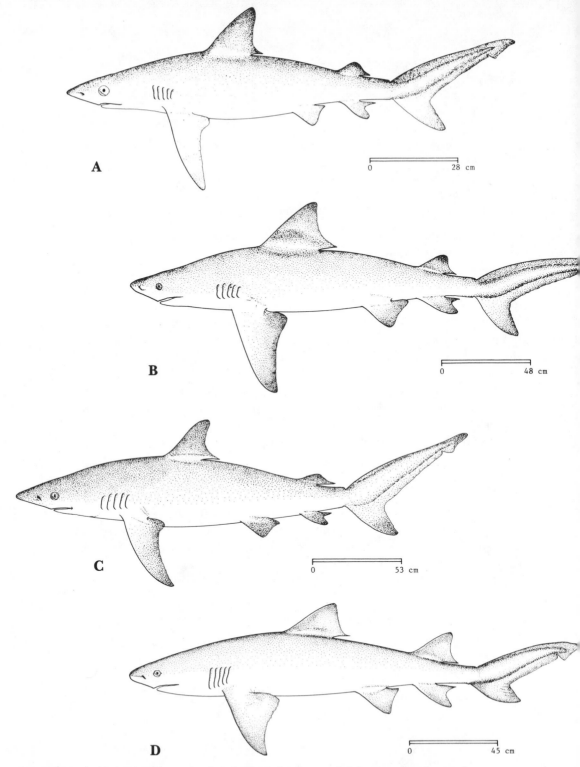

Gray shark: The look-alike *a*, reef; *b*, bull; *c*, dusky; and *d*, lemon sharks can be distinguished by differences in tooth shapes and small external anatomical details that are not easy to see in the water. *(Illustrations courtesy of Fish and Agriculture Organization)*

The bonnethead and hammerhead sharks go to extremes at the front end. Their heads are spade-shaped, with the eyes set far out on the edge of what looks like a diving plane. Odd as this shape seems, no one has shown what advantage it confers. However, a large **great hammerhead,** *Sphyrna mokarran,* emerging out of the gloom is good enough reason to leave the water and contemplate the strange shape from dry ground. This shark has been implicated in enough human attacks to warrant prudence.

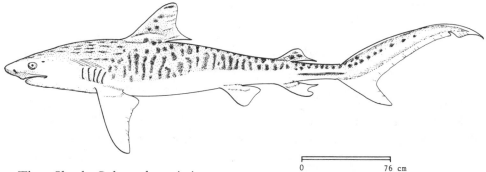

Tiger Shark, *Galeocerdo cuvieri*
(Illustration courtesy of Fish and Agriculture Organization—the United Nations)

The typical coloration of a shark is blue, brown, gray, slate, or black on top (dorsally) blending to off-white on bottom (ventrally): This is commonly called *countershading.* A few species are mottled or, like the **tiger shark,** *Galeocerdo cuvieri,* barred vertically with dark stripes on a lighter background. Tiger sharks are members of the *requiem* family, a number of whom have the same general shape and color and have been lumped by fishermen as "gray sharks." These include the **reef shark,** *Carcharhinus springeri,* the **bull shark,** *C. leucas,* the **dusky,** *C. obscurus,* and the **lemon shark,** *Negaprion brevirostris.* By far, the tiger is the most threatening. It can get to be three to five meters long and will eat almost anything. Although usually sluggish, when aroused by food it is unpredictable and dangerous. It swims pelagic and coastal waters from Cape Cod to the Antilles, the Gulf of Mexico, and the southern coast of California.

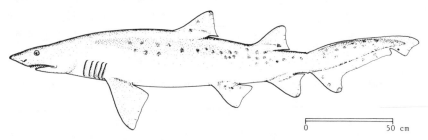

Sand tiger shark, *Odontaspis taurus*
(Illustration courtesy of Fish and Agriculture Organization—the United Nations)

Divers on tropical reefs have discovered that sharks need not continually swim to stay alive. The **nurse shark,** *Ginglymostoma cirratum,* often rests with its head tucked under a coral ledge. So do the bullhead and lemon sharks. Dr. Eugenie Clark, of the University of Maryland, investigated similar instances of "sleeping" sharks in underwater caves off the Yucatan coast. All were "gray" sharks or the easily recognized blue shark. During this "sleep," the sharks' gills actively pumped water. They could be moved gently, but became wide awake if handled roughly.

Sharks sense their surroundings by sight, olefaction, and sound. The eye of a shark, unlike those of bony fish, has a reflector in it to enhance sight in very dim light. In the rear of the eye, reflecting plates bounce incoming illumination back through the light intensity detectors (rods), nearly doubling their efficiency. Their eyes also contain cones, light detectors that discriminate specific sets of wavelengths, which suggests they see colors. Their preference for "yum yum yellow" in artificially stimulated attacks on commonly used water objects suggests the traditional Mae West color may not be the most appropriate waterwear, but, in real life, attacks on divers in black wet suits suggests that dressing like a seal isn't wise either.

Sight is the shark's guiding sense when closing in on prey. Sound and olefaction detect the prey at a distance. *Olefaction* is neither the direct equivalent of smell or taste but sort of a protein-trail detector. Judging by the size of the shark's olefactory sacs and its responsiveness to long chum slicks, it's put to good use. Paired sacs connect to the nostrils and provide directional cues.

To detect sound, vibrations propagating through water, sharks, like bony fish, have a lateral-line sensing system. A series of interconnected canals filled with mucus have periodic external pore openings and form a network over the head and down each side of the body. The lateral line detects low-frequency vibration. The fluttering of a wounded fish, about twenty-five cycles per second, will draw sharks from several hundred meters away.

Sharks have another sensory system that remains a puzzle to scientists. Pores on the head not connected with the lateral line, called the *Ampullae of Lorenzi,* are said to detect depth changes, slight temperature and salinity changes, and slight electrical-field gradients generated by another animal's muscle activity.

When mariners talk about the unpredictability of sharks, their concern centers upon when and what it chooses to eat. The shark's eclec-

The thin line between adventure and foolhardiness is for each of us to draw. *(Photo by D. Nelson)*

tic tastes, precision biting apparatus, and capricious decisions about what constitutes opportunity gives pause and doubt as to their intentions.

The contents of a shark's stomach doesn't correlate with its readiness to feed. Sharks with empty stomachs have passed up easy handouts; others, already stuffed, have opted for more. A shark's appetite is generally overexaggerated. In captivity they eat between three and fourteen percent of their body weight weekly, but often go for long periods without eating at all. One juvenile tiger shark at the Miami Seaquarium went five months without eating, then ate forty pounds of blue runner in thirty-six hours. Low water temperature may halve normal feeding; so does breeding. Some sharks won't feed at all during mating season or when close to *pupping*, the proper term for sharks giving live birth.

Their feeding history can be judged by the size of their livers. Sharks do not have fatty body tissue. Their reserves are stored entirely in their livers. Its two great lobes can occupy most of the body cavity and amount to twenty-five percent of body weight or shrink to less than three percent after fasting.

Mating styles among shark vary with body flexibility. The male **spiny dogfish,** *Squalus acanthias,* wraps himself around the female. Large, stiffer sharks mate side-by-side. All fertilization takes place internally, but the development of the egg can be external or internal. Some sharks lay an egg in a tough, chitinous case, like the *mermaid's purse* of the skate. (If you find one intact under water, keep it moist and cool and take it to your aquarium. Carefully cut open one end of the case and, with luck, out will slide a thrashing embryo attached to a yolk sac. You can raise it to "birth" and beyond.)

The egg can also develop internally, unconnected to the mother, relying on its own yolk sac, or, as with the **sand tiger shark,** *Odontaspis taurus,* the first embryo will eat subsequent eggs passing down the uterus. With requiems, hammerheads, and some dogfishes, the egg is attached to the mother through placental-like tissue.

Swimming with Sharks

Diving with sharks calls for preparation, the advice and accompaniment of experienced and competent companions, prudence, and a modicum of fortitude. Avoid diving with the foolhardy and avoid foolhardiness yourself. Learn to identify sharks. Recognize the differences between the known harmless ones and those with a bad reputation. Be aware of conditions that can bring trouble. The potential hazards increase with the shark's size and degree of stimulation.

The reef sharks of Florida, the Caribbean, and California, the many "gray" or "ground" sharks, the tiger shark, the lemon shark, and the great hammerhead are the commonest encountered in warm shallow water. The gray and lemon sharks usually just swim by for a look, reluctant to approach too closely. If a fish has been speared or there is chum in the water, the sounds and odors will draw sharks in. You may suddenly find yourself confronted with several big creatures closing rapidly. If the shark initially mistakes you for the source of the sound or odor, it will usually turn away as it sees the real prey. If it gets too close, face it, shout at it, and make a sudden aggressive move by extending your arms out toward it: Divers claim that, if timed right, the technique never fails. Hans Hass, a pioneer diver of the 1950s, carried a closed black umbrella with large eyes and a ferocious set of jaws studded with big teeth painted on it. When charged, he pointed it and abruptly opened it at the attacking shark. He claimed that this, too, never failed to intimidate. But if nothing works, ward the shark off with whatever you have in your hand. Avoid being brushed by it unless you are covered up; its skin is remarkably abrasive.

Large reef sharks, the three-to-five-meter tiger or great hammerhead, rarely charge, but close in by circling slowly. Use the opportunity to move toward the boat, land, a reef, or other divers. Remember that you cannot outswim a shark, so move away with as little disturbance as you can muster.

The pelagic blue shark will rarely approach you directly unless drawn in by food, while the oceanic whitetip and **silky shark,** *Carcharhinus falciformis,* will swim straight for you. The white shark is regarded as the most dangerous to man, although, in the West Indies, the tiger shark is in the running for that distinction. Incidents occur in the tropics more frequently than in temperate waters, although a recent series of attacks on divers in the colder waters of California, where sea otters are common prey, suggests the white makes no distinction between them and human swimmers.

In short, dealing with such powerful creatures on their own terms requires knowledge, self-control, and common sense. Don't stay in the water with big sharks. Never go in water where sharks are actively feeding. Don't grab or poke at sharks or block their paths in narrow confines. And don't dive alone. Sharks are less likely to attack if two divers are present.

ODD FISH

Keep a weather eye on the surface of warm seawater when you can, for your reward may be a glimpse of a shark basking, a sea turtle, or an **ocean sunfish,** *Mola mola.* Your first sighting of an ocean sunfish will engender total disbelief. It looks like a huge fish cut in half and lying on one side; the tail fin is nothing more than a short, thick flap of skin.

Ocean sunfish one to two meters in length stray into mid-Atlantic coastal waters in summer and fall. They are common the year around

Though big, the ocean sunfish, *Mola mola*, is harmless and often wantonly destroyed.

in tropic seas. They breed at sea and feed on drifting jellyfish, comb-jellies, salp, crustaceans, and young fish.

You can bring a boat close to them, get in the water, and swim over to them without fear. As you approach, they will abandon the horizontal position and swim vertically, watching you. You can try to hitch a ride by grabbing the tail fin, but that almost always sends them swimming downward, and a flick of their tail instantly dislodges you. Little or nothing is known of their life history or their wanderings.

Skittering away from the track of your boat, flying fish enhance any sea journey. Their long pectoral fins, nearly half body length, when extended let them glide for twenty to one hundred meters. However, they seldom rise more than a meter or two off the sea surface. If the sea is millpond-quiet, you will see the trace of the sculling motions they make with the lower lobes of their tail fins on the water's surface. More than twenty-five species exist.

Warm water will also bring a few visitors around an anchored boat. The **banded rudderfish,** *Seriola zonata,* native to the Gulf of Mexico

A pair of banded rudderfish, *Seriola zonata,* hover under the keel of an anchored boat.

and the south Atlantic, will align with the keel or anchor line. **Pilotfish,** *Naucrates ductor,* are also attracted to boats and, occasionally, a **remora,** *Remora remora,* will attach itself to the hull.

The life histories of the remoras are a mystery. Their fry have been found in the open Atlantic in June and July. Common in the West Indies, the adults attach to sharks and turtles. The suction disk on the head of a remora is powerful. Torres Straits fishermen catch turtles by tying a rope to the fish and paying out the line until it latches onto a turtle, then haul in both.

FISH WATCHING

Getting close enough to fish to watch their behavior, either in open water or in shallows, takes patience and adherence to a simple yet seldom-heeded set of rules.

First, think about yourself. Are you properly weighted and dressed? You normally need more lead for trim in oceanic water than in coastal water. Ballast yourself to be slightly heavy; about a kilo excess will do. That allows you to free-dive effortlessly and hover over the bottom without floating up, yet will not hinder your return to the surface.

Protect your body. If you plan to spend time near the bottom or around protruding walls of rock or coral, be sure to cover yourself to avoid needless scrapes and cuts. A wet suit serves well in chilly water but may be too warm in the tropics. In warm water, use a sweat suit or wash-and-wear pants and a long-sleeved shirt and garden gloves. You may not appear in the height of fashion, but you will save yourself from torn skin and prevent a debilitating case of sunburn. The snowbird just off the plane from the North has a safe sun tolerance only of half an hour. It's easy to spend four or five hours in the water the first day and realize you have been deep-fried after it's too late.

Before you plunge in, stop and consider yourself from the point of view of the fish. You are in a great hurry to get going. Once in, will you flail down and proceed away full bore? What do you think will be the natural response of a fish of reasonable size at the approach of a big noisy behemoth that is over two meters long from head to toe? Exactly—you would get out of its way, and the fish will, too. Not far, just enough to feel out of harm's way. The ocean is full of so many threats that a safe distance is seldom more than ten meters away. As your noise subsides or blends into the background of other sounds, the initial threat will be forgotten. The fish will reappear to look you over or go about its business.

However, where spearfishing is practiced, large gamefish won't let you within their sight or return, if they see you, until after you have left. You must do your looking in places where fish are not hunted. Even so, certain species are naturally skittish, and all you may get of them is a fleeting glance.

All species are wary of anything as large or larger than they are that stares at them. If you appear preoccupied with something other than them, their curiosity may compel them to investigate *you*. The longer you stay in one place, the more you will see the indigenous life. In the tropics, reef fish that hid at your initial approach will reappear, and passersby will detour over to you to see what you are up to. Barracuda are notorious for appearing over your shoulder, as if to figure out what you find so interesting. Parrotfish, goatfish, and hog-fish will deliberately alter course to pass near you. They won't stay; they just look and go.

The lesson? Stay put and let the fish come to you. If you must get closer, swim slowly and stay close to the bottom. Fish feel threatened by anything overhead. Avoid quick movements—they will evoke a flight response. Don't approach a fish head-on; move in at an angle, occasionally stopping or slowing down. You can often maneuver close to a large grouper or ray this way. To get close to a school of fish, swim parallel with them, edging in from the rear. If you approach them from the front, they will turn away from you.

Certain sounds attract some fish. Try tapping one rock on another or scratching two rocks together. Some will flee, but others approach. Gentle thuds seem to arouse interest, while high-pitched sounds invariably repel.

Food is a surefire attractant, but its use has drawbacks, both to yourself and the fish. If you make repeated visits to the same place, its permanent residents will note your arrival and anticipate an immediate handout. I've seen a regularly fed grouper weighing in excess of a hundred kilos rush over to his benefactor, eat what was offered, then pester him for the rest of the dive. The grouper had become vulnerable to any human that wanted to turn him into a fish-fry, and the humans around him had no opportunity to see him conduct his life normally.

Bait probably is a reasonably acceptable way to draw a cryptic invertebrate from its hiding place deep in a crevice. A little chopped clam fed in with a basting syringe may draw out a creature you otherwise would have little chance of seeing. Nevertheless, use bait sparingly. It can alter the behavior of ocean creatures to a point where you can't tell induced reactions from natural ones.

3

At Sea: Whales, Seals, and Turtles

INFREQUENT VISITORS ALONG MUCH OF OUR COASTLINE, WHALES, dolphins, seals, sea lions, sea otters, and turtles draw a crowd wherever they appear. Naturalists travel great distances to see them, even though they will not be able to spend enough time among them to gain more than a nodding acquaintance with their ways of life.

Yet naturalists do make contributions to our understanding of these creatures, uncovering bits and pieces of their mysterious life histories. Your encounters may be by choice or chance. Since you never know what these meetings may bring to light, the more you know about them beforehand, the better the likelihood you can add to your own storehouse of experience and recognize the uniqueness of an event or a mode of behavior not well documented as yet.

WHALES

Over the last several decades, the whale has become the animal embodiment of the open sea. No other sea creature has such a strong grip on the imagination. People who rarely venture on the oceans have come to sympathize with the plight of both the leviathans brought to near-extinction by overhunting and the great losses of dolphins and porpoises drowned in drift nets and dying from disease.

The toothed whales and the whalebone, or baleen, whales comprise the order Cetacea. Toothed whales include the dolphins and porpoises, the rarely seen beaked whales, and the sperm whales. Most of the dolphins are marine, but a few live exclusively in freshwater rivers.

In the baleen whales, the teeth have been replaced by horny plates emerging from the roof of the mouth. The plates are frayed along the edges and act as coarse sieves. Baleen whales either gulp water and food and, by closing their mouths expel the water, or skim the water for food, letting it accumulate on the plates.

The fin, humpback, sei, minke, and Bryde's whales are gulpers, preferring fish and shrimplike euphasiids. The right and gray whales are skimmers, mainly feeding on amphipods. The gray whale often feeds close to the bottom, and in so doing, leaves large shallow depressions in soft bottoms.

All are warm-blooded mammals who bear live young and suckle them for a time. They range in size from a meter to thirty meters in length, and are the largest creatures on earth. Ten species of baleen whales and sixty-six species of toothed whales are known, of which forty-three species are dolphins and porpoises.

A sperm whale that stranded on Fire Island, New York, and was subsequently treated for pneumonia with penicillin, swims back out to sea. *(Photo by Richard Ellis)*

Big whales, twelve to twenty-six meters (forty to eighty-five feet) in length, are not easy to identify in the water; so much of them lies below the surface. Nine species swim the North Atlantic and Pacific. You can tell them apart by regional distribution, profile, the shape of the spout, diving form, and, if you can get close enough, features that are indistinct at a distance.

Don't wait until you are at sea to begin learning what to look for. Know at least how to sort out the main differences before you go. Looking it up on a rolling ship with a pair of binoculars in one hand while hanging on with the other is hardly the time or place to begin from scratch.

Unless you visit Arctic waters, you can eliminate the **bowhead,** *Balaena mysticetus.* It only frequents the outskirts of pack ice. Should you get there, though, you can recognize it by its two-part "Loch Ness Monster" profile: a raised hump followed by a low, smooth portion of its back.

The **blue whale,** *Balaenoptera musculus;* the **fin,** *B. physalus;* the **sei,** *B. borealis;* **Bryde's whale,** *B. edeni;* the **humpback,** *Megaptera novaeangliae;* and the **sperm whale,** *Physeter catodon,* all possess a dorsal fin you can see either as the whale lies in the water or as it begins a dive. Neither the **right whale,** *Eubalaena glacialis,* nor the **gray whale,** *Eschrichtius robustus,* have a dorsal fin.

Humpback whale off Cape Cod *(Photo by Don Riepe)*

Individual humpback whales can be distinguished by their flukes. This is "Salt," a female, who has turned up off Stellwagen Bank for the past eleven years and is probably the most looked-at whale outside of captivity. *(Photo courtesy of Cetacean Research Center, Provincetown, Maine)*

The "boxcar" profile of the sperm whale clearly distinguishes it from other members of this group who from a distance in a choppy sea look remarkably alike.

The blue has a smaller dorsal fin than the sei or Bryde's whales. It is set back near the tail and is not immediately exposed upon surfacing. Getting closer, the blue has a broad U-shaped head. The fin, with whom the blue is often confused, has a narrow, V-shaped head.

The fin whale has a single raised ridge on its head, as does the sei. Bryde's whale has three head ridges, which in some are not prominent enough to allow it to be distinguished from the sei. However, Bryde's whale is seldom found north of Virginia on the East Coast or north of Central California on the West Coast.

Watch the back of the whale as it starts a dive. The blue, the humpback, and the sperm whales thrust their flukes clear of the water as they begin; the fin, the sei, and Bryde's whales do not. (These six all have a dorsal fin that will show as they begin their dives).

The right whale lifts its flukes high out of the water as it starts a dive; the gray's flukes barely clear the surface. Neither has a dorsal fin.

Right whales are scarce: less than a thousand still survive worldwide and only three hundred fifty swim the Atlantic. Although rare, they regularly show up over Jeffrey's Ledge and Stellwagen Bank off the coast of Massachusetts every spring. They calve off the coast of Georgia and northern Florida in winter, then migrate north to their feeding grounds.

The gray whale is common in the Pacific but was hunted into extinction in the Atlantic by the late 1700s.

On the East Coast the fin is the most commonly sighted large whale; the humpback is a close second, in part because we know where to look. From May to October they, too, frequent Jeffrey's Ledge and Stellwagen Bank. The world population of humpbacks is nearly ten thousand; there are at least ten times that many fin whales. Only seventeen hundred humpback whales roamed the western North Atlantic in 1989, but their migration routes are so regular that we can predict who, when, and how many will likely show up at one of the favorite feeding or calving grounds. Swimming both the Atlantic and Pacific coasts, they mate and calve in the south in winter and spend summers feeding in colder northern waters.

Humpbacks migrate northward in the spring from their major calving grounds on the Navidad, Silver, and Mouchier banks near the Dominican Republic, and grounds off Puerto Rico. In March and April, some arrive for a brief stay off Bermuda and are within small-boat range. Their "songs" were recorded there by Roger Payne who, with tape recorder and hydrophone, worked from a inflatable runabout. Proceeding north, five distinct groups head either for Greenland, Iceland, Labrador, the Gulf of St. Lawrence, or the Gulf of Maine to feed.

The humpback's head is flat and covered with large knobs. The body is black on top, white below, and often covered with large patches of barnacles. The long, bumpy, and ragged flippers are nearly a third of total body length. When one dives, it raises its midsection into a high rounded arch and raises its flukes clear of the water.

If you can get a photograph of the flukes reasonably perpendicular to the camera, you can identify the individual whale. The fluke markings and scalloping are as distinctive as fingerprints, and researchers have catalogued thousands of them.

Make notes on what the whales are doing: date, time, location, numbers, state of the sea, and whatever else you can glean. Keep it in a log. When you accumulate enough information and pictures, there are several scientific groups on your coast who will be glad to get your data. Contact any of the whale-watching groups given in the Chapter Notes.

Humpbacks are active and acrobatic. They will leap clear of the water, then fall on their backs or sides, creating a booming shower of spray. They will raise their flukes and bring them down flat with a resounding smack. They will roll on their sides and slap the water with their long flippers or swim straight up out of the water, relax, and settle down again without splashing.

The leaps of the humpback caught the fancy of the press back around the turn of the century. Rarely a summer passed without an eyewitness feature story vividly describing a sea battle between a whale and a swordfish. The swordfish was never visible in these fights of imagination, but it seemed to be the only rational explanation for the whale's inexplicable behavior.

A leaping humpback propels thirty-two metric tons aloft, no mean feat of strength. Sometimes it lands in a belly flop, but more often it does a midair twist and lands on its back. It usually breaches in sequences of an average of nine leaps, one every forty seconds or so. One whale, who must have been going for a personal record, breached 130 times in seventy-five minutes on Silver Bank, off Hispaniola.

Breaching seems more frequent among social whales—humpback, right, gray, and sperm whales. Once breaching starts up in a pod of individuals, it continues, spreading among the members. Just what message it conveys is still being puzzled over by cetologists, but the whales know a good time when they are having one.

By comparison, the fin whale (often called the finback) is relatively sedate. Not much of its enormous bulk normally shows above water. Reaching a length of twenty-four meters, its dark gray back is distinctly ridged toward the tail, but is rarely as mottled or scarred as the sei or blue whale.

The way the fin whale surfaces and dives distinguishes it from other species. Coming up, the fin whale emerges head-first, barely breaking the surface, blows a cone-shaped mist five meters high, and rolls forward, arching its back, exposing a small dorsal fin. When sounding, the body continues to roll forward without exposing the flukes.

The sei whale, with whom the fin can be confused, swims with both its head and dorsal fin simultaneously exposed, and on sounding, doesn't arch its back as much as the fin does. It slides in, barely baring its back beyond its dorsal fin.

What attracts the humpback and fin whales, and **Atlantic white-sided dolphins,** *Lagenorhynchus acutus,* to Stellwagen Bank and Jeffrey's Ledge are fish, **sand lance,** *Ammodytes americanus,* in particular. Slender, no more than eighteen centimeters long, these fish congregate in dense shoals, providing fodder not only for whales, but also for cod, haddock, halibut, salmon, mackerel, striped bass, and bluefish. The sand lance rarely ventures over rocky or muddy bottoms, perhaps because it can burrow into sand so quickly. It seems to swim into it, digging with its sharp nose and propelled by an eel-like body motion.

The gray whale has become the center of interest for the California coastal whale-watcher. It can be spotted from shore while migrating along the coast, or at breeding grounds near the mouth of the Gulf of California.

The mature gray whale is just over twelve meters long. Traveling singly, in pairs, and in small groups (large pods are rare), they can swim a steady four knots and average one hundred kilometers a day when migrating. Like most northern-hemisphere baleen whales, the gray whale spends June through September in the rich waters of the Arctic. In fall they migrate southward, appearing off the California coast in late November, and arrive at their breeding grounds, the bays along Mexico's Baja Peninsula, in December and January. There they calve and breed. The female matures in her eighth year, then breeds every other year. She carries the calf for one year and nurses it for seven months. At the breeding grounds, females without calves attract males who, in twos and threes, vie for her attention. Often, several males can be seen lying alongside a female, pushing and shoving each other for a favorable position. The northward migration of mothers and calves begins in March.

Gray whales can be recognized easily. Their head is sloped and has a forehead bump. They have no dorsal fin, but about where the dorsal fin would be on a fin whale the gray has a series of nine to fourteen bumps that extend dorsally to the base of the flukes. The basic color is gray, but white where the whale has been scraped. Grays are so mottled and scratched that they give the appearance of being gray on white. Barnacles cover the upper jaw, the sides of the head, and the flippers.

Spyhopping gray whales in Scammon's Lagoon, Baja, California *(Photo by Sally Dick)*

If you venture among these whales, keep in mind that a mother is exceptionally protective of her calf and will be vengeful should the calf be injured or endangered. She will pound a boat to pieces with her flukes or overturn it if provoked into a rage.

Until 1970, what we knew about big whales was mainly gleaned from carcasses hauled in by commercial whalers. Today, individual whales can be identified by their natural markings and scars, which allows the researcher to track the animals over wide expanses and long periods of time. Since many whales travel in groups, their social lives, breeding histories, and longevity can be ascertained from sightings and photo identification.

Fluke marks alone will identify humpback whales. Nearly four thousand individuals have been catalogued this way in the Atlantic, and two thousand in the Pacific. Body markings—natural color-pattern variations, callosities, barnacle incrustations, and scars—are now used for gray, fin, blue, right, sperm, Bryde's, minke, and killer whales. The technique has been applied to dolphins and pilot whales, as well.

Cataloguing and comparing new and old photographs has revealed a great deal about whales. Members of small social groups, especially mothers and young, stay together for many years, as much as a generation. Pods of killer whales and sperm whales are remarkably stable, and whole genealogies have been worked out for some.

Every now and then, the chance stranding of a whale that can be saved and released has allowed an opportunity to attach radio transmitters to it that will track where it goes and what it does. UHF transmitters coupled to a receiver aboard a TIROS-N weather satellite have allowed whales to be followed day and night over huge expanses for three months at a time, the life of the transmitter battery.

In June 1987, a tagged pilot whale, released off Cape Cod, quickly rejoined a pod of a hundred companions, swam eighty kilometers a day, and made repeated dives that ranged in duration from six seconds to nine minutes. During the night, it made especially deep dives, presumably to feed on squid.

Among the medium-sized whales, the **minke,** *Balaenoptera acutorostrata,* and the **Atlantic pilot whale,** *Globicephala melaena* on the East Coast, and the **short-finned pilot whale,** *G. macrorhynchus,* on both coasts, are also frequently seen inshore.

A mature minke averages seven meters (twenty-three feet) in length. It has a narrow head and a prominent and highly recurved dorsal fin. Basically black or dark gray on top, you can just see the light gray of its undersides at the water line.

On the East Coast, the Atlantic pilot whale ranges from North Carolina north, and the short-finned pilot whale ranges from North Carolina south. Both are all-black, have a bulbous head, and a low dorsal fin with a long base. They travel in small groups of four to six, or herds of fifty or more.

The **killer whale,** *Orcinus orca,* is also medium-size and is unlikely to be confused with anything else in the sea. Its chunky body, high sail-like dorsal fin, and large white areas on a jet-black background make it distinctive. It travels in small groups and in larger pods of twenty-five to thirty.

Dolphins and porpoises are also toothed whales. The terms "dolphin" and "porpoise" are often used interchangeably by fishermen, but there is a difference: Most dolphins have a beaked snout; porpoises do not have such a snout.

The **harbor porpoise,** *Phocoena phocoena,* is the smallest, about 1.5 meters in length and fifty kilograms in weight when mature. Once common inshore in the northern temperate waters of both coasts, their numbers have dwindled. Fisherman dislike them because they feed on cod, whiting, mackerel, and herring. They will circle a school of herring, tightening the frightened fish into a ball, then rush them to feed. They avoid ships and rarely jump clear of the water. Little is known of their habits. Pairs may mate for life, but this supposition remains scientifically unproven.

The **Atlantic bottlenose dolphin,** *Tursiops truncatus,* is rarely seen

Wild bottle-nosed dolphins have been known to strike up friendships with humans, returning to the same waters day after day to play. *(Photo by David Caldwell)*

more than eight hundred kilometers (five hundred miles) from shore. It has a limited home range, surprising for an animal that lives twenty-five years and can swim at fifteen knots.

Scientists are equivocal about the longevity of whales and dolphins in the wild. The blue whale is said to live over twenty-five years. Captive animals live longer than those in the wild, who succumb earlier not only to the vagaries of an uncertain existence, but to simple wear and tear. However, a few longevity records exist. A dolphin nicknamed "Pelorus Jack" took to following ships off the coast of New Zealand and continued to do so for thirty-two years before vanishing. How old he was when he started, no one knows.

Whale Watching

Commercial fishermen and charterboat captains sight whales at sea fairly regularly off both the East and West coasts. A sea voyage of more than a few days without dolphins riding the bow waves for at least part of the time would be a disappointing trip indeed. A totally unexpected breeching of a large whale close by the boat is a once-in-a-lifetime experience for the part-time seafarer.

In the last ten years, the popularity of whale watching has enticed skippers of large party boats into running one-day to one-week whale cruises. These trips are seasonal. Humpbacks and fin whales are sought by boats from Montauk, Long Island, to York, Maine, from May to October. On the West Coast, boat cruises are available to Scammon's Lagoon in Baja, California, and other traditional breeding grounds of the gray whale. Day trips to see grays and other species run from Seattle to San Diego. You can book these trips in advance with natural history or environmental groups, or directly with the charter boat.

Whale watching on Stellwagen Bank has become a local growth industry in Provincetown, Massachusetts, with ten or more boats making two trips a day. Large party boats are sea-kindly but take foul-weather gear aboard even on a sunny day for protection against spray (and take medication before you leave if you are prone to seasickness). Bring binoculars. If you don't see whales, look for birds and don't despair. A day at sea is never a waste of time.

On the East Coast, whale sightings, other than those on the fishing banks of New England, most frequently occur near the edge of the Continental Shelf. If you have the means to get there often and can get within a kilometer of the whales, you can listen to them. To do so, you need to know something about electronics (or have a friend who does). You need a hydrophone, an "impedance matching" device (which is often built into the hydrophone), thirty or more meters of cable, a preamplifier, amplifier, and tape recorder (which may all be obtained in a single unit), earphones, buoys, and a length of shock cord.

To avoid surface noise, rig the hydrophone so that it dangles from a buoy ten to fifteen meters below the surface. Create a loop of slack in the cable with a piece of shock cord to lessen noise that would otherwise be created by waves jolting the hydrophone. Naturally, the boat and all aboard must be quiet during recording.

A surfacing humpback whale blows a balloon-
shaped vapor cloud.

BLOW BY BLOW

Most whale watchers do their sightseeing from the deck of a tour boat dedicated to that purpose. Usually there is a naturalist on board who can identify whales and tell you something about their natural history.

It's all well and good to stand passively by, taking it all in, but much more rewarding if you can be the first to spot the whales and make an educated guess as to species.

You can see the "blow" long before the whale itself. The shape, height, and frequency of the blow of each species is relatively distinctive. For example, the humpback blow is a three-meter-diameter balloon of vapor, four to eight times at fifteen- to thirty-second intervals after a long dive. The fin's blow is an elongated ellipse, to six meters in height, three- to seven times at intervals up to several minutes each. The minke, a much smaller whale, has a low, indistinct blow.

On the West Coast, the gray whale blows a bushy cloud three to four meters high. Seen from a fore-and-aft view, the blow is actually two columns, creating a heart-shaped effect.

Learning the characteristic shapes of all the whales will take a lot of sea time and many encounters, so watch for whales from a distance as well as when you are up close.

A stereo cassette tape recorder can be modified to accept the hydrophone input on one channel and your verbal comments on the other, allowing you to match up sound with sighted action. Remember to protect your electronics from the sea: Salt water is remarkably corrosive and a few spritzes of sea spray will ruin it.

Analyzing the sound requires equipment that filters out unwanted frequencies and transforms inaudible and audible frequencies into a graphic record. All this is well within the ability of an electronics buff.

You will not be able to record the entire sound spectrum of some cetaceans because it can range in frequency from fifty cycles per second to two hundred kilocycles with a dynamic range more than one hundred decibels (dB), but a high gain (60–80 dB) amplifier with a broad, flat frequency response in the audible range will catch enough to make the project worthwhile. (For do-it-yourself advice on how to build a hydrophone, see the references in the Chapter Notes.)

SEALS AND SUCH

Aside from Cetacea, other orders of seagoing mammals live in our coastal waters: the seals, sea lions and sea elephants, the manatees, and the sea otters.

Except for the manatee, all prefer cold water. On the West Coast, you can see **Steller's sea lion,** *Eumetopias jubatus,* the largest of the eared seals; the **California sea lion,** *Zalophus californianus,* the most common trained seal in circuses; the **sea elephant,** *Mirounga augustirostris,* whose preposterous nose immediately identifies the male of the species; and the **harbor seal,** *Phoca vitulina.*

Harbor seals were once widespread along the Northeast Coast, but their numbers have dwindled. The **gray seal,** *Halichoerus grypus,* occasionally visits the coast of Maine, but its domain lies farther north. The gray and harbor seals are sometimes confused, but the gray has a much longer snout. In the Canadian Maritime Provinces, it is called "horse-head."

The **sea otter,** *Enhydra lutris,* once ranged from Baja to the Bering Straits, but was nearly exterminated by sealers for its pelt. It is making a comeback along the central Californian coast. This time it is protected and welcomed for its proclivity for making a meal of the sea urchin, who, in runaway numbers, nearly denuded coastal kelp

The California sea lion breeds along the southern Californian and Mexican coasts but ranges as far north as Vancouver Island *(Photo by Sally Dick)*

A baby harbor seal hiding in the rocks on San Martin Island, Baja, California. *(Photo by Sally Dick)*

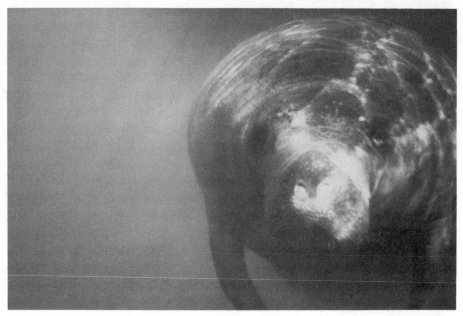

(Illustration by Sally Dick)

THE MYTHIC MERMAID

Meet the manatee, an air-breathing mammal that weighs in excess of 450 kilos, measures three meters from head to tail, and is gentle as a lamb. These lumbering herbivores feed on water hyacinth in Florida's fresh waters during the winter months and spend their summers browsing on marine grasses in offshore beds.

An endangered species, the total population of *Trichechus manatus* is probably less than two thousand. Nearly eighty-five percent have been scarred by encounters with power boats, which are the major cause of their decreasing numbers.

Manatees, often called sea cows, can be spotted along all Florida's coastline as well as along the gulf states. One favorite congregation point for them is Crystal River in Florida, from January through March. Warm (22° C) spring water attracts them, as does the warm effluent of a nearby nuclear power plant.

If you sight a manatee, record where, when, how many, doing what, and send the information to the National Fish and Wildlife Laboratory, 2820 E. University Avenue, Gainsville, FL 32601. If you sight a tagged Manatee (the tag floats), get the color combination of the tag and phone the Lab at (904) 372-2571.

If you encounter an injured Manatee or deliberate harassment of one, which is illegal, phone 1 (800) DIAL-FMP.

beds in the absence of the otters. No creature appears more at leisure than a sea otter reclined on its back, slowly munching an urchin or just dozing among the gently swaying kelp fronds.

All these animals are *amphibious,* able to live on land *and* water, and haul out on shore to bask and pup. Keep your distance from them in breeding and pupping season.

SEA TURTLES

A chance meeting with a turtle at sea happens frequently enough to warrant knowing something about it. Creatures of warm water, a few species make northward excursions up the Atlantic Seaboard in summer.

Five species visit United States shores. **Kemp's ridley,** *Lepidochelys kempii,* and the **hawksbill,** *Eretmochelys imbricata,* seldom venture north of Florida. The **green turtle,** *Chelonia mydas,* the **loggerhead,** *Caretta caretta,* and the **leatherback,** *Dermochelys coriacea,* favor tropic waters, but venture into midlatitude temperate waters in summer. A sixth, the **olive ridley,** *Lepidochelys olivacea,* tracks across the Caribbean and south Atlantic Ocean, but is rarely ever seen north of Mexico.

A hawksbill turtle heads for the sea after being hauled ashore by an inquisitive diver.

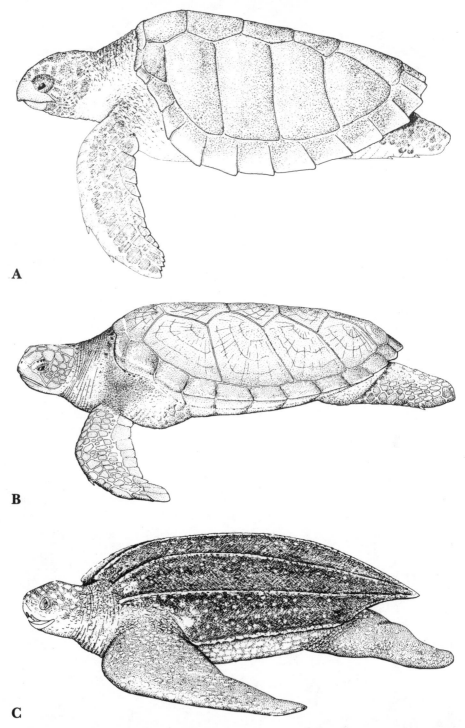

A

B

C

Of the big turtles that roam the oceans, the *c*, leatherback is the largest and most distinctive. The *a*, loggerhead; and *b*, green turtles are not easy to tell apart at a distance. *(Illustration courtesy of Fish and Agriculture Organization—the United Nations)*

Of the five, the leatherback is the largest, often 1.5 to 1.7 meters in length and a half a ton in weight. Its back is so distinctively different from the others that you can tell it at a glance. Five to seven longitudinal ridges run along a back that is not shell, but tough leathery skin.

The leatherback is pelagic and, except for nesting, spends its entire life on the high seas, living on a diet of jellyfish. Rarely resting, it wanders long distances. It dives frequently, averaging a depth of sixty meters (two hundred feet) and a stay of twelve minutes. Dives as deep as a thousand meters have been recorded. Just what these turtles are after, no one knows, but it's speculated they are hunting jellyfish in the deep *scattering layer,* a zone where drifting animal life concentrates.

Leatherback nesting grounds—sandy beaches adjacent to deep water—are scattered worldwide, but shoreline development is slowly eliminating them.

The shell of a turtle consists of fused, bony plates: a main series down the back called *vertebral scutes,* an adjacent series down each side called *costal scutes,* and smaller ones ringing the shell, called *marginal scutes.* The plate of fused bone on the underside is called the *plastron.* Between the plastron and the top shell, some species have a series of plates along their sides called the *inframarginal scutes.*

If the shell is heavily moss-backed, look at the turtle's head. The green turtle has two prefrontal scales between its eyes, the loggerhead two pairs. Young loggerheads have a very bumpy keel down the middle of the shell, which disappears as they grow older. The shell of the green turtle is smooth irrespective of age.

The green turtle and the loggerhead are about the same size, just under a meter long at maturity. If the shell is reasonably clean, count the costal scutes; the green has four on each side, the Loggerhead five or more.

Mature green turtles are vegetarians, and therefore often spotted in shoal water among marine grasses. The loggerhead will feed on grasses but also eats shellfish, conch, goose barnacles, sponges, and jellyfish, including the Portuguese man of war.

The swimming speed of a green turtle is impressive. An adult female caught and brought to the Bermuda Biological Station was docile enough while begin weighed, measured, and tagged, but the instant the sling containing her was winched out and lowered into the water, she turned her engines on full-bore. A full running torpedo couldn't have shot out the chute much faster. She headed directly for open sea and passed out of sight without ever surfacing.

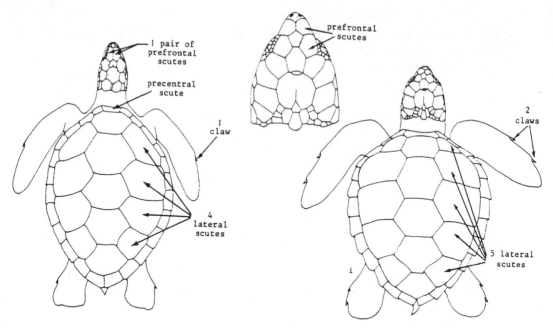

The main external differences between the green turtle *left,* and the loggerhead *right,* are simple enough to discern up close unless their shells are heavily covered with algae. *(Illustration courtesy of Fish and Agriculture Organizaion— the United Nations)*

Green turtles weigh an average of 113 to 182 kilograms (250 to 400 pounds), while Kemp's ridley and the hawksbill rarely exceed 36 to 64 kilos (80 to 140 pounds). The hawksbill is easy enough to identify; its scutes overlap like shingles, and the edge of its shell is serrated. The ridley can be confused with the loggerhead. Both have five or more costal scutes, but on the underside the ridley has pores in its inframarginal scutes. ridleys are a smaller species than the loggerhead. The head is more arrow-shaped and the flippers smoother.

The Kemp's ridley is fast passing from the scene, more so than the other turtles on the endangered list, who are doing poorly in their battle for survival. Sea turtles have a tragic flaw in their life styles. They must return to land to breed and are programmed to come back to the beaches from which they hatched. Kemp's ridley is down to one breeding beach, Rancho Nuevo in Mexico, a few hundred kilometers south of Brownsville, Texas. Kemp's ridleys all come ashore at

Mask, fins, and snorkel (and practice) can take you safely down to full fathom five to see a (mainly) natural, unchanged world.

A foureye butterfly fish, *Chaetodon capistratus,* swims on Molasses Reef in the Florida Keys in 1966. Lush scenes like this are fast disappearing in the Keys as human encroachment takes its toll.

Almost any hard floating object that has been at sea for a time in warm waters will develop a colony of goose barnacles, *Lepas anatifera*.

The blackfish, *Tautoga onitis*, varies in color from a mottled mouse brown to solid black as it grows older.

A typical New England stony bottom; the hard surfaces, living and inanimate, are covered with an encrusting red algae and occupied by the purple sea urchin, *Arbacia punctulata*.

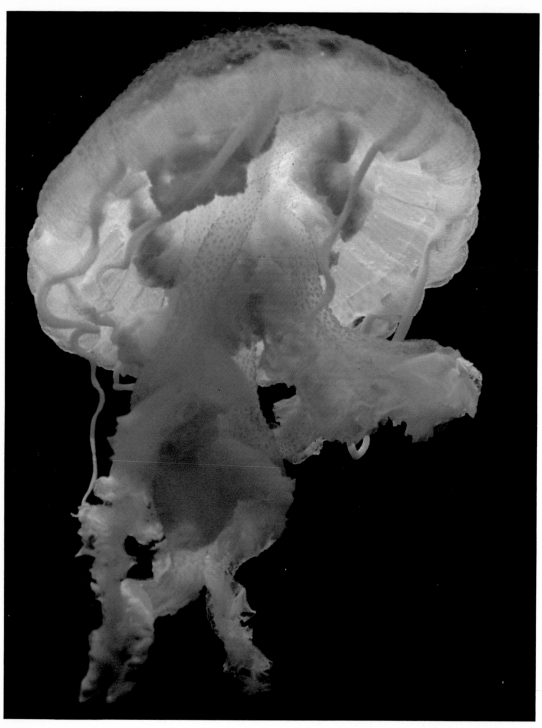

This pretty little jellyfish, *Pelagia noctiluca,* can be found all along the Atlantic coast from Florida to Cape Cod. It is luminescent at night.

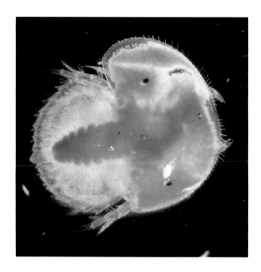

A horseshoe crab, *Limulus polyphemus,* one day after hatching. With every subsequent molt it will increase in size and the tail will become more prominent.

The greater amberjack, *Seriola dumerili.* Compare this photo with a line drawing of this fish in an identification manual and note how many of its fins fold away while it swims.

The spiny lobster, *Panulirus argus,* usually holes up during the day, foraging at night. These lobsters have been seen making long migrations, single file, in deep water; from where-to-where and why isn't known.

The up side of the upside-down jellyfish, *Cassiopeia xamachana,* which settles to the bottom with its downside up.

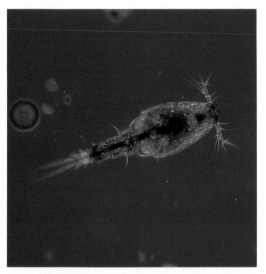

Copepods are the most numerous multi-celled animals in the sea. The round object near its tail is a diatom, *Cosinodiscus.*

The sea slug, *Hypselodoris zebra,* is a nudibranch that frequents tropical grassy shallows and feeds on sponges.

A view of the business end of the Portuguese man of war, *Physalia pelagica.* Its tentacles can extend downward 45 meters in a quiet sea, invisibly combing the waters for prey.

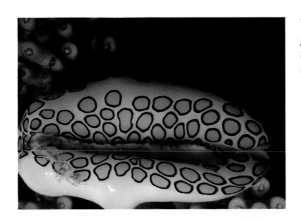

The flamingo tongue snail, *Cyphoma gibbosum,* feeds on sea fans. The brilliant colors are in its mantle, not its shell, so enjoy it alive (ie. don't collect it).

The humpback whale, *Megaptera novaeangliae,* is noted for its spectacular breaching which seems to be a form of play. When one starts, others invariably follow. —photo credit Wayne Mossman

Large numbers of the jellyfish, *Cyanea capillata,* often show up in Atlantic coastal waters in spring. To a human its sting is irritating but not painful.

The moon jelly, *Aurella aurita,* drifts by. The four-leafed clover rings are reproductive tissue.

The spotted sea hare, *Aplysia dactylomela,* releases a harmless purple fluid when disturbed. It normally spends its day browsing on algae.

A fringed filefish, *Monacanthus ciliatus,* in the company of several young banded rudderfish, *Seriola zonata.*

A parrotfish holds still while blueheads, *Thalassoma bifasciatum,* pick off parasites at a cleaning station. Only the supermale has a blue head; most blueheads are yellow.

A young cunner, *Tautogolabus adspersus*, is dwarfed by two anemone, *Metridium senile*.

French grunt, *Haemulon flavolineatum*, are common on reefs.

The colonial ascidian, *Clavelina picta,* will grow on a variety
of surfaces including this sea rod. The colony develops
by asexual budding.

The blue angelfish is obvious but do you see the lobster, *Scyllanides,* clinging to the rocky coral? I didn't until the photos came back from the photofinisher!

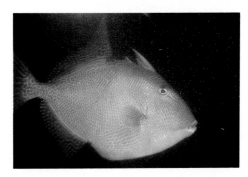

It's a triggerfish but neither fits the color description of the ocean triggerfish nor the fin description of the rough or gray triggerfish. Unless you can find an expert or get further information you may have to settle for a family name rather than pinning it down to genus and species.

The sea rod, *Plexaurella nutans,* fully
extends its polyps when feeding.
If disturbed the polyps withdraw
into slits in the rod.

the same time; the *arribada* is well known to the locals. There they suffer human poaching as well as animal predation.

At sea, thousands of ridleys and young green turtles drown in shrimp nets. The National Marine Fisheries Service has designed and tested a *Turtle Excluder Device* (TED), which, attached between the mouth and cod end of the net, deflects the turtle to an escape hatch. Louisiana shrimpers are refusing to use it, saying it is too expensive (it costs $150 to $400), loses shrimp (a few escape with the turtle), is inconvenient, and is a government intrusion on a traditional fishery. The alternative is to reduce towing times to less than ninety minutes, which the turtle can survive.

Every creature takes its chances against the perils that await it, but the turtle has had more than its share of man-made troubles. The adults show up like clockwork at traditional feeding grounds. Green turtles graze on grasses on the Miskito Cays and Mosquito Bank, Costa Rica, regularly enough to support a fishery hundreds of years old. Their breeding beaches are well known: from Quintana Roo, Mexico, to south Brevard County, Florida. The breeding beaches of the hawksbill in Yucatan, Campeche, and Mona Island, and the loggerhead along our Southeastern Seaboard and the coast of Mexico are also well known.

Green turtles take thirty years to reach maturity, then return seasonally to nest at their natal beaches. They lay 60 to 160 eggs in a *clutch,* burying them nearly a meter down in soft sand. They may lay as many as seven clutches over a two-week interval. The eggs take two months to hatch.

Swimming green turtle *(Photo by George Huey)*

The little turtles all emerge from the egg within a few days and collectively face the problem of extricating themselves from their underground nursery. They do this by what Archie Carr, the late world-reknowned turtle expert, called "witless cooperation." Simply by periodically flailing, some knock the sand off the top of their tomb and others below displace it to the bottom. Thus, the turtles and the cavity they occupy rise inexorably to the surface. Once there, they head for water. Crabs and gulls take an immediate toll, as does misdirection from land-based lights.

One of the last mysteries of the sea turtle's life history was its whereabouts between hatching and maturity. As it turns out, we have created a new set of hazards at that stage as well.

Hatchlings immediately make their way to open sea. Feeding on plankton, they are drawn to oceanic convergences, lines of downwelling water. These drift lines are relatively stable and collect rafts of seaweed and driftwood, providing attaching places for the larvae of sedentary animals, feeding grounds for pelagic crabs, and shelter for small fish.

Sargasso weed, with its own indigenous floating communities, is especially providential to the young turtles, supplying both food and shelter. The turtles remain on the floating sargasso-weed islands as passive migrants for three to five years, depending on the species. Unfortunately, these drift lines have also become collection points for floating marine debris, particularly tar and plastic. The little turtles

Kemp's ridley *(Photo courtesy of Center for Environmental Education, Washington, D.C.)*

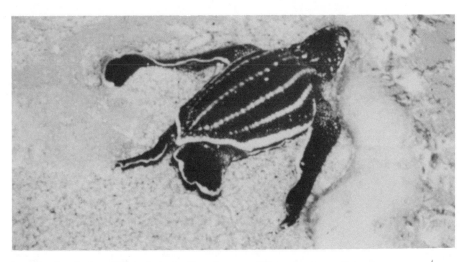

Leatherback hatchling heads for the sea *(Photo by Karen Eckert)*

make the mistake of feeding on inedible morsels with mortal consequences. Thousands of little loggerheads have washed ashore on the Florida coast with stomachs full of tar pellets and plastic beads.

Nearing maturity, some species give up their earlier carnivorous habits and head for the grasses in shallow water. During that migration, green and loggerhead turtles feed on jellyfish, as they also do on the way to breeding grounds. The leatherback alone remains at sea.

Again the turtles run afoul of mankind's negligence. They mistake plastic for jellyfish or, like oceanic birds and fish, get entangled in lost or discarded nets and cordage. The reason for the deaths of most of the large turtles found floating or washed ashore is ingested plastic, however. Plastic scrap at sea has grown so prolific that downwellings are often outlined in white: the accumulation of styrofoam flotsam.

4

The Drifters

❧

THE BULK OF LIFE IN THE SEA IS NEITHER THE LEVIATHANS nor the teeming swimmers, but the drifting life. Some are so small you need an electron microscope to make out their structures; others are easy to see with the unaided eye. These drifters make up over ninety percent of all sea life. The plants among them contribute between twenty and thirty percent of the world's oxygen supply.

This drifting life is collectively called *plankton,* a word derived from a Greek stem meaning wanderer, that has been applied to all marine forms who go where the currents take them. Some can swim, others can rise and fall in the water column by changing buoyancy, but none have the ability to come and go as they choose.

Lumping this hodgepodge of life under one sobriquet was the doing of Victor Hensen, a professor of biology at Kiel University in Germany, in the late nineteenth century. This enormous diversity of sea life had been discovered earlier by another German, Johannes Müller, who, in 1845, first collected samples of it by towing a fine-meshed net.

The full extent of this hitherto unknown richness really hit home with the voyage of the *Challenger.* Between 1872 and 1877, HMS *Challenger* tracked around the world, collecting samples of life from the tropics to polar seas, from shallow waters to the deeps, returning home with a cornucopia of new species. The reports of this expedition filled over fifty volumes and unleashed a worldwide interest in marine plankton. In its wake, one expedition after another trekked the world's oceans, each adding to the knowledge of the sea's seldom seen little creatures.

Pelagic life, that is, life in the open sea, can be found both in *neritic* waters (those waters over continental shelves) and *oceanic* waters (those over ocean deeps). Neritic waters hold most of the ocean's abundance because they are richer in nutrients than are oceanic waters. However, where deep, nutrient-rich ocean waters surface at an upwelling, the open sea can also teem with life.

Plankton provide the base of the food web in the sea. *Phytoplankton,* simple plantlike organisms, manufacture the raw materials upon which the *zooplankton,* small animal life, feed. Zooplankton, in turn, are the prey of fishes.

The size range of plankton is enormous: from viruses and bacteria to giant salps and jellyfish with bell diameters to two meters (six feet) and tentacles that can extend twenty meters (over sixty feet).

PHYTOPLANKTON

Phytoplankters are the primary producers of the sea, playing a role akin to grasses and plants on land. Like land plants, they manufacture living matter from the energy of the sun, carbon dioxide, and nutrients by photosynthesis. Unlike the plants on land, almost all the phytoplankton are microscopic algae, either one-celled or multi-celled chain forms. And also unlike land plants, they process carbon dioxide from the water, not the air.

Larger plants have not adapted well to the open sea. One exception, **sargassum weed,** drifts in the mid-Atlantic. However, it evolved from land-attached coastal predecessors.

At sea, smallness has advantages. Tiny creatures have far more surface area for a given volume than do larger ones. This increases their resistance to sinking and lets them take in nutrients and dissolved gases simply by absorbing them.

The algae of the sea consist of *diatoms, dinoflagellates, coccolithophores,* and to a lesser extent, *blue-green* and *green algae* and *cryptomonads.* Diatoms build and encase themselves in an intricate framework of silica

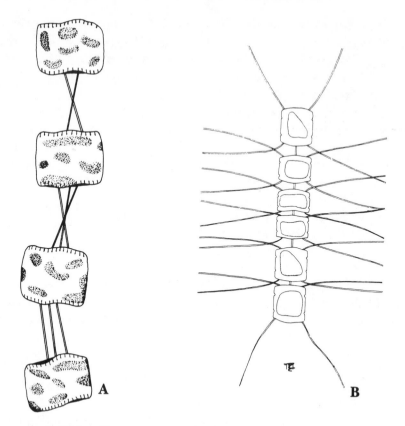

Two typical chain diatoms: *a, Cyclotella* sp. (15 mm diameter); and *b, Chaetoceros* sp. (5 mm diameter) *(Illustration courtesy of Florida Department of Natural Resources, "Hourglass")*

filled with pores through which they conduct their metabolic affairs. Dinoflagellates enclose themselves in cellulose and possess two whip-like structures that provide movement.

The miniscule coccolithophores look like balls covered with round plates. These plates (*coccoliths*) are made of calcium carbonate. Coccolithophores can be tremendously abundant. Satellite photographs have spotted blooms of them extending over vast expanses of ocean.

Blue-green and green algae and cryptomonads, very small pigmented cells with a single flagellum, a whip-like extension that can actually propel the organism, are most common in estuarine waters. Some are the size of bacteria. Just how they fit into the main food web of the sea is still a matter of scientific conjecture.

All algae contain chlorophyll that is carried in pigment bodies called *chloroplasts*. These are the centers that manufacture carbohydrates from carbon dioxide, water, nutrients, and sunshine. Because sunshine is rapidly absorbed by seawater, most photosynthetic activity goes on in the top 15 meters (about 50 feet) of the sea, although in clear oceanic water, photosynthesis will take place in waters to 100 meters (325 feet) deep. The rate of photosynthesis depends not only on nutrient supply but on the total number of photosynthesizers. The denser the concentrations of phytoplankters, the less light will penetrate beneath the shading by the uppermost creatures.

Diatoms are the most important food supply for the zooplankton. In the midlatitude neritic waters, their numbers swell tremendously in spring and fall. The spring outburst, the larger of the two, usually occurs in March and early April. Surface waters, rich in nutrients collected over the winter, need only the lengthening days and some warming to trigger their explosive growth. They are grazed down by zooplankton during the summer, falling in numbers not only because of predation but because the surface waters run short of nutrients.

In fall, the surface water cools and is overturned by rough seas. Storms mix the barren top water with the richer bottom waters. The diatoms briefly increase again, then dwindle as the winter light fails.

A few of the diatom species eaten by mussels: *a, Navicula* sp.; *b, Biddulphia* sp. *(Illustration courtesy of U.S. Bureau of Fisheries)*

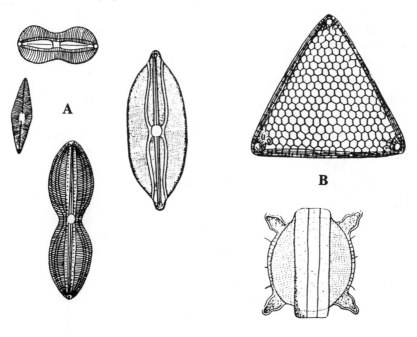

The productivity of temperate water in summer is over twenty times that of the yearly average for tropic water. Phytoplankton growth in midocean tropic seas rarely converts more than 0.05 grams of carbon per cubic meter of seawater per day into living matter. In early summer in Long Island Sound, the figure is near 1.2 grams of carbon per cubic meter per day. The low output of tropical water is caused by the shortage of "fixed" nitrogen, that is, nitrogen bound to other elements that can be assimilated by the cells to make amino acids and proteins essential to all life.

Unlike plant life on land, where legumes "fix" nitrogen from the air, only a few species of blue-green algae can do so in the sea. Most of the nitrogen in waters over the continental shelves comes from land runoff and from reprocessing what has sunk to the bottom. In open sea, sinking animals and plants carry nitrogen down to the deeps where it is lost to the upper layers.

Surface sea life depends on the surplus carbon production of phytoplankton. Each species of phytoplankton does best at its own optimum light level; thus, as depth increases, primary production rises until it reaches maximum value, then falls off as the light fails. So there exists a zone in which the net production of food is positive and below which it is negative.

Photosynthesis produces oxygen. During the day, phytoplankters increase its supply. During the night, both phytoplankton and zooplankton *respire,* consuming oxygen and releasing carbon dioxide. Using a simple dissolved-oxygen kit, if you sample inshore water shortly before dawn and again in the afternoon, you can measure this difference and gain an understanding of this activity.

You can collect phytoplankton (and zooplankton) with a net of fine-meshed nylon. The very smallest creatures will escape through the meshes, but most diatoms and dinoflagellates will be retained by a 150 to 200 mesh-per-inch cloth. The cone-shaped net is held open at its wide end by a metal hoop that is attached to a tow rope by a bridle. The narrow end is closed off by a collecting jar made of metal or plastic. The towing must be done slowly (no faster than one to two knots) or else the water will simply be pushed aside by the back pressure of the net and will not be filtered through it. Towing time depends on the number and size of the creatures in a given volume of water. At times, a few minutes' tow will gather more than enough. Offshore in open sea or in tropic waters, long tows may be needed to harvest the sparse inhabitants.

If you pour the catch into a clear jar, you can easily see pinhead-sized zooplankton zipping about in the water. If your sample contains a preponderance of phytoplankton, you may not be able to make out the individuals, but the water will appear colored, usually yellow or brownish.

Under a microscope that provides twenty-five to one-hundred diameters magnification you can identify most of the common forms. Some may be too small to tell much about. To get an idea of relative size, put a human hair on the microscope slide with your sample. Its diameter is about forty micrometers (one and a half thousandths of an inch). Anything smaller than that diameter falls in the size category of the *nanoplankton* (five to sixty micrometers) or *ultraplankton* (less than five micrometers), and you will not be able to identify them: coccolithophores, smaller flagellates, some single-celled blue-green and green algae, and bacteria included.

Diatoms and many dinoflagellates are large enough to be seen in detail at one hundred diameters magnification. Diatoms are easy to spot by their unique casings. Some are round, others elongated. The round ones have a pill-box shape from which spines or hairlike projections may protrude. Often they are attached by silica threads, forming long chains.

To reproduce, the pill-box separates into two halves, called *valves* (the whole box is called a *frustule*), and the separate valves form new partner valves to complete the frustule. This leaves one of the pairs the same size and the other a trifle smaller. Thus you will find that diatoms of the same species come in a variety of sizes. Obviously, this process can't go on too far or the species would shrink to nothing. After suffering some diminution, the small cell revolts, casts off both valves, swells, and rebuilds anew.

The silica skeletons of diatoms are marvels of elaborate repetitive design; often so small and intricate that resolving them with an optical microscope is impossible and requires the capabilities of a scanning electron microscope. Indeed, in the early days of optical microscopy, specially prepared specimens of certain species of diatoms were used to test the quality of the lenses.

The dinoflagellates are mostly single-celled and can move by beating one or both flagella in such a way that they are propelled forward, often rotating as they go. Their body shapes range from simple ovals to ornate plated or spined configurations. Replete with hooks, knobs, and polygonal plates, they look like objects from the age of

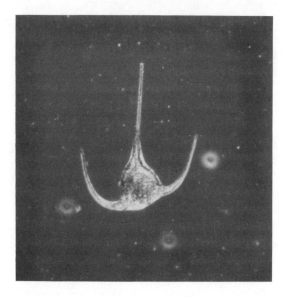

Ceratium tripos, a dinoflagellate

armor. Some are so spiky (*Ceratium,* for example) that they have few predators.

Some are phosphorescent. In temperate seas, *Noctiluca* is one such, common in summer, and is an added joy to a night trip to sea. .Pass your hand through the dark water and watch the myriad tiny sparkling lights they emit.

The dinoflagellates are well known for their sudden and enormous reproductive surges, their very numbers discoloring the waters over wide areas. If green or blue, their growth is scarcely noticed, but if red, shore communities will speak of "red tide," which often means local shellfish beds will have to close. Clams, oysters, mussels, and other shellfish filter them out with impunity but store the toxin the dinoflagellates have manufactured. If fish or man eat those shellfish, serious poisoning can result: *paralytic shellfish poisoning* (PSP), which, at the least, is unpleasant, and fatal at the worst.

ZOOPLANKTON

All the major animal phyla and most of the minor ones can be found in the plankton either as adults or in a larval stage. Some will spend their whole lives adrift (*holoplankton*), while others will pass through a

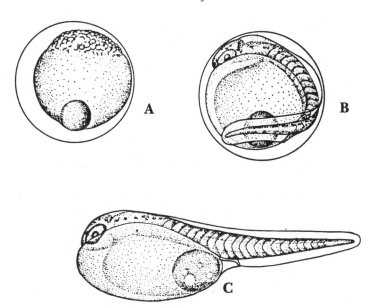

Planktonic stages of the bluefish: *a,* egg at five hours after fertilization; *b,* embryo at forty-five hours; *c,* newly hatched larva (length 2.2 mm). *(Illustration courtesy of Fish and Wildlife Service)*

drifting stage and go on either to be strong swimmers or surface settlers (*meroplankton*). The egg of a fish develops through a drifting larval stage and eventually into a swimmer; the larva of a barnacle transforms into a stage that finds a hard surface and settles down to a sedentary existence for the rest of its days.

Small single-celled animals may turn up in your tow net (although a lot will pass through all but the finest meshes). Members of the Protoctista, the most likely will be the *tintinnids,* tiny cup-shaped creatures whose rim is often lined with a beating crown of cilia. They are diatom eaters and, being transparent, you can see the ingested diatoms in their guts under a microscope.

If you take your tow near a bottom swept by fast currents, you may find *foraminiferans* among the catch. Many of them have coiled shells made of calcium carbonate, much like a snail except far smaller. Some ocean bottoms are covered with so many of their settled shells that the sediments are called *foraminiferan ooze.* The best known and most abundant foram, *Globigerina,* is common in warm seas.

So too are the *radiolarians*, who build a latticed exterior scaffolding of silica. Their shells have settled out in vast numbers on some tropic sea bottoms, forming a sediment called *radiolarian ooze*.

Sponges lead attached, sedentary existences, and would hardly be expected to be found afloat. They do, however, shed a drifting form, an asexual larva called a *gemmule*, but you will seldom see it in net hauls. To see it at all, collect a few pieces of sponge from a wharf piling or rocks during late July or August and put them in an unaerated jar of cool sea water. You should have some free-swimming gemmules within hours. Don't wait too long to examine them. They settle down to a quiet attached life within days.

All three classes of Cnidaria, jellyfishes and kin, occur in the plankton, although the *anthozoans*—anemones and the like—are limited to their planula larvae, a small, ciliated, free-swimming, oval-shaped form. The *syphozoans,* true jellyfishes, are there as both adults and larva. Larvae just released from polyps can often be reared to more advanced larval forms, but rarely to adulthood.

Small jellyfish kept in a dish with other members of the plankton go on fishing for them as though they were still at sea. You can watch them capture small prey. If you place a tentacle under a microscope and brush it with a bit of protein, the sting cells will discharge. You will be able to see the threadlike hypodermic apparatus that delivers a paralyzing toxin to the prey.

In winter and early spring along the North Atlantic Coast, drifts *Cyanea,* a formidable jellyfish whose tentacles can extend six meters (twenty feet) or more. They are often surrounded by pods of small fish, young pollock and butterfish in particular, who manage to survive the barbs of this floating fortress without ill effect. If you swim toward the jellyfish, the little fish will crowd to the side, away from you. If you persist and come closer, they abandon their haven and head for another jellyfish.

The clear blue water of the Gulf Stream carries with it many exotic species of Cnidaria that have no common names—*Diphyopsis, Physophora, Cupulita*—and some well-known, such as the Portuguese man of war. You will see long, fragile creatures with arrowhead or grape-cluster floats, pie-shaped and bowl-shaped pulsing balls, and odd rectangular forms with structures like clear boxes within boxes. They are all large enough to see easily and are best seen in the water. To swim from a boat with mask and snorkel can provide you with a visual treat that exists nowhere but the sea. However, remember that

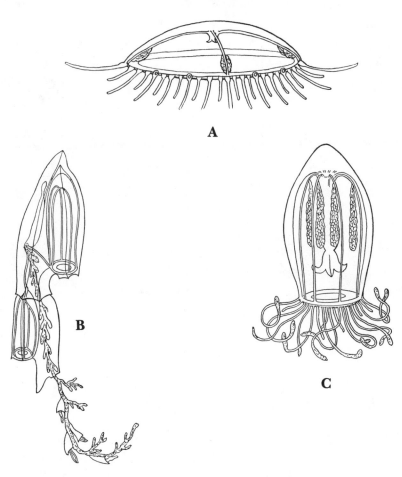

A

B

C

Pelagic jellyfish: *a, Obelia; b, Diphyes bipartita; c, Aglantha digital*
(Illustration courtesy of U.S. Bureau of Fisheries)

the more tropical the waters, the wiser it is to keep your skin covered. The sight of *Physalia*, the man of war, or *Chrysaora*, the **sea nettle,** is not sufficient compensation for the bolt of lightning you will feel if you are unlucky enough to blunder into either one with bare flesh.

Like the jellyfishes, and for many years lumped with them, the comb-jellies, Ctenophora, are common sights in the plankton. In temperate waters, their numbers swell in summer and you may find your net clogged with their jellied remains. *Beroe, Mnemiopsis,* and the smaller *Pleurobrachia* seem to fill the sea during the hottest months.

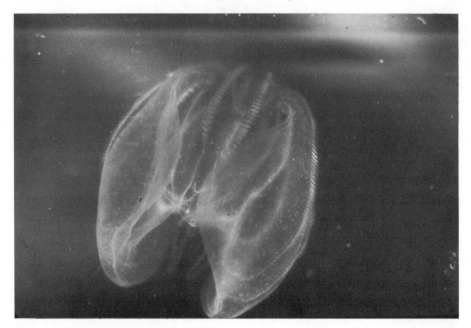

The comb-jelly, *Mnemiopsis leidyi*

Less common, but a spectacular sight, is *Folia,* a colorless, beltlike (and belt-sized) ctenophore often called **Venus girdle.**

Ctenophores will remain alive for a few days in cool, aerated sea-water. There you can watch the diamondlike flashes of light that refract off their beating comb-plates when struck by sunlight. And you can watch them flash by night as well—they are phosphorescent. Simply disturb them by gently stirring the water (without tearing them apart!) and they will pulse with a bright glow.

The most prolific of all the animals you will find in your plankton net are the *copepods,* crustaceans who are members of the phylum Arthropoda. They range in size from a few tenths of a millimeter to two millimeters. Dozens of species may turn up in your catch. If you do repeated tows over the seasons, you may find the dominant species change as the year progresses, depending on where you tow.

Copepods are the single most important source of animal food for small fishes. Vast schools of herring and mackerel count on their endless numbers and continually roam the sea looking for waters rich with them.

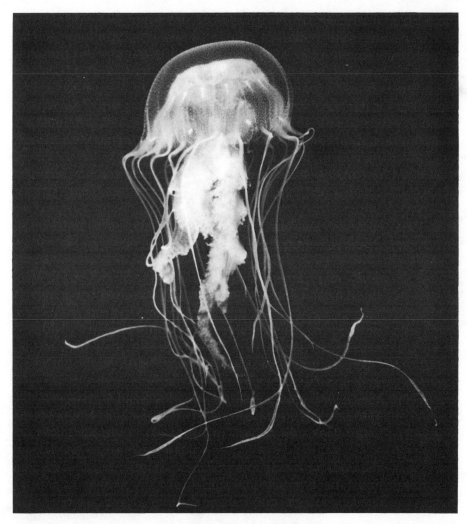

The sea nettle, *Chrysaora quinquecirrha (Photo by M. J. Reber)*

The Crustacea are a huge class, encompassing thousands of species in the plankton: mysids, shrimp, ostracods, cladocerans, an endless array of variations on the central arthropodian scheme. And not just the adults; the form of larval crustaceans often differs considerably from that of the parents.

Like so many other invertebrates, many of the adult crustacea do not live in the plankton; but their larva do: shrimp, crab, and lobster

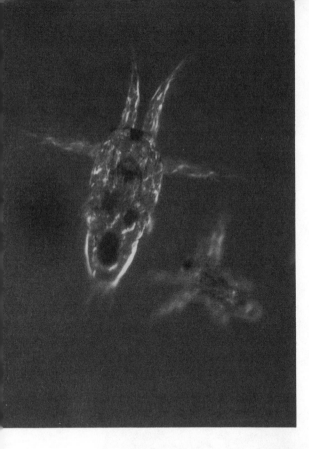

Nauplii larvae of copepods

The cladoceran *Evadne* is common in western Atlantic waters.

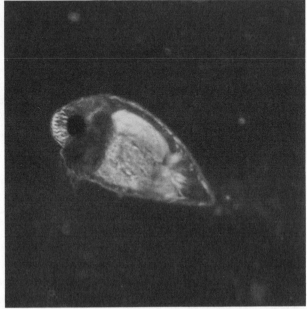

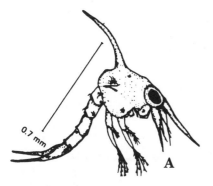

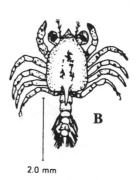

Planktonic stages of the rock crab: *a,* early zoea; *b,* megalops. *(Illustration courtesy of National Marine Fisheries Service)*

among them. Shrimp and crab start out as spiny *zoea* and transform into a *megalops,* a sort of small, bug-eyed, compressed version of the adult. Copepods have a *nauplius* stage, as do acorn barnacles, which precedes adulthood. Sorting all the forms out, especially as to species, involves paying attention to the details of appendages and body segments. Even so, the young of some remain enigmatic and take a practiced eye to recognize.

Nauplius larva of a barnacle

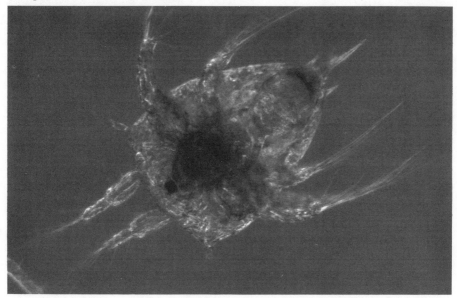

The plankton net, especially in summer, will take in the young of many others in the animal kingdom: bristled polychaete larva, petaled *veligers* of snails, tiny valves of clams, the *bipinnaria* and spined *echinopluteus* of starfishes and sea urchins, and the eggs and fry of fish turn up.

As intimidating as this brief list is, you can get to know the major drifters after a relatively short acquaintance. You needn't try to nail everything down by species; simply spot critters by family, order, or class. Few biologists know a majority of the planktonic species by sight. If you find plankton interesting and worth your time, it won't be long before you surpass the knowledge of all but the dedicated specialist.

Examples of lesser-known phyla will also turn up regularly. The arrow-worms, *chaetognaths,* have a "prefish" appearance: slender, symmetrically finned, with a head equipped with spines that can snap shut on prey like ice tongs.

Using a microscope on a boat is difficult. Boat motion sloshes the specimen around and motor vibration jiggles the image. However, try

An arrow worm on the prowl. The jawlike bristles shut on contact.

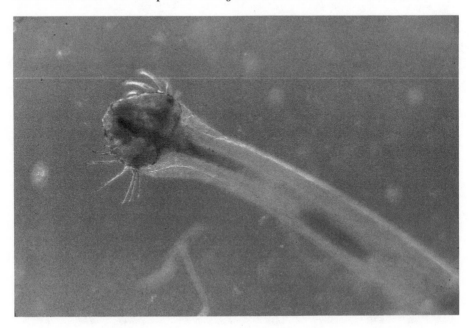

to view your plankton catch as soon after you capture it as you can. Because of loss of oxygen or overcrowding, many of its numbers will expire within hours. Alive and well, their hurly-burly activities will give you some insight into how they live in open water. Copepods dart from one place to the next, pause momentarily, then dash on. Others are in constant motion, their swimming forms dependent on their means of propulsion: some move in a straight line, some pulsate, others wind along in a corkscrew fashion.

You can preserve your sample with formalin. Mix it in a five percent solution with seawater. Be aware that delicate forms often contract violently in formalin and require pretreating with a narcotizing agent. Also, because of recent fears surrounding the safety of *formaldehyde,* the active ingredient in formalin, it may be hard to obtain.

A good substitute that will also serve as a narcotizing agent is alcohol. You can use either vodka or rubbing alcohol. Hundred-proof vodka is fifty percent ethyl alcohol. Rubbing alcohol is seventy percent isopropyl alcohol. To narcotize the catch, add alcohol to the seawater suspension slowly, drop by drop, until signs of activity cease. Usually, the plankton will settle to the bottom of the container. Pour off most of the seawater and slowly add alcohol until you have a fifty-fifty solution. Within a few days, pour off the water-alcohol solution and replace it with straight alcohol. This will preserve the plankton for years.

PLANKTONIC DISTRIBUTION

Plankton in the sea are not uniformly distributed. Thousands of square miles of ocean are relatively barren of plankton; coastal areas and upwellings are relatively abundant. One reason for this is food supply: vital nutrients are not evenly distributed.

Continental-shelf water receives minerals and organic matter from freshwater runoff and recovers the nutrients from debris (such as dying plankton) that sinks to the bottom. Upwellings raise up rich bottom water from the deeps and support immense swarms of plankton. These upwellings occur along the western shorelines of continental margins abutted by deep trenches or swept by enormous surface currents. Offshore winds displace the surface water away from the coast and bottom water rises up the steep continental slope to take its place. The source of these cold, rich waters is the polar seas, where the winter waters, made dense by cooling and saltier by the removal of fresh water in the form of ice, sink and begin a long underwater journey toward the equator.

Phytoplankton are enormously more abundant than zooplankton. In a cubic meter of seawater, a productive, coastal-temperate patch of water might hold five thousand to fifty thousand zooplankters, fifty thousand to one million large phytoplankters, and ten million small phytoplankters. In open ocean these numbers might be from a hundredfold to a thousandfold less.

However, plankton occurrence is patchy, even in productive seas. Within a few kilometers you might find an abundance of a single species, and nothing of it in adjacent waters. You can occasionally observe this without resorting to a microscope. In clear tropical waters it is not uncommon to see a parade of jellyfish passing by like so many rolling dinner plates, hundreds at a time. The whole panoply may last five minutes, and then you may not see another of that species again for weeks.

The kinds and number of plankton in a given area change with the seasons. The source of these changes is the sun, whose energy drives the whole food web in the sea. Because of the tilt of the earth's axis relative to its orbit around the sun, both the length of the day and the angle sun rays strike the water change with the time of year. This affects water temperature as well as the light intensity and duration in the water column.

The extremes of lightness and darkness occur at the poles. There, temperature ranges from −2° C to 5° C (28° F to 41° F) over the year, and the light varies from total darkness to continuous, albeit low intensity, light. Summertime activity is enormous. First, the phytoplankton peak, then zooplankton do. The pace is feverish enough to outstrip the nutrient supply in the Arctic, but not in the Antarctic.

At the other extreme, from the Tropic of Cancer past the equator to the Tropic of Capricorn, the sea temperatures are a constant 26° C to 28° C (79° F to 84° F), and the length of the days is twelve hours, plus or minus one hour, the year around. Light penetrates into clear seawater because the sun is high in the sky, yet planktonic life is rarely as abundant as in temperate water. The water is nutritionally poor.

What tropical planktonic life lacks in numbers it makes up for in diversity; many more species are present but each are fewer in number. According to one marine biologist: "Of the roughly five hundred species of copepods, eighty-seven percent are found in the warm regions, while only three percent are to be found in the cold northern regions, and only six percent in the southern." This holds true for other animal groups in the plankton as well.

Tropical species are more ornate than their temperate cousins. For example, the tropical copepods tend to have long and frilled append-ages. These adaptations are thought to help prevent sinking in the warmer, less-dense seas.

Staying suspended without propulsive power has been solved by a number of adaptations. Becoming lighter than water is one answer. Certain jellyfish do this with large gas-filled floats, the Portuguese man of war and the **by-the-wind sailor,** *Velella,* for instance. Other smaller cnidarians use tiny floats (*pneumatophores*) that hold their bodies close to the surface. To see them, rig the mouth of your plankton net to a float in such a way that it skims the water's surface. Plankton at the interface of the sea and the air are different enough from that below to have been given a special name: the *neuston.*

Diatoms use two strategies to stay up: They produce oil globules that reduce their overall body density closer to that of seawater, and they have hairlike projections and a flattened body shape, which in-creases their frictional resistance to sinking.

The midlatitudes, 20° to 50° north and south, show greater changes in seasonal plankton abundance as one proceeds poleward. Sea tem-peratures vary by 10° C to 15° C (18° F to 27° F) from summer to winter and, at 50° latitude, light energy at the surface decreases by a factor of ten in winter.

Warm continental-shelf water in the midlatitudes supports an enor-mous amount of life in spring, summer, and fall. In winter, plankton are scarce. Many diatoms and dinoflagellates lapse into resting phases; some float, some settle on the bottom. Nutrient levels rise as winter storms mix surface and bottom water. As spring approaches, the water warms and days lengthen. In late February or early March, a diatom bloom begins and continues well into late spring. A *thermocline* (a sharp demarcation zone between the warmer surface water and the colder bottom water) forms in coastal water. In the deeper ocean, upper waters also warm, but the temperature gradient over depth is more gradual than inshore.

Zooplankton, feeding on phytoplankton, increase rapidly. Phy-toplankton begin a steep decline; their numbers are cut by predation and a rapidly dwindling supply of the nitrates and phosphates neces-

sary to sustain their rapid reproduction. By the end of summer, zooplankton have peaked. Then they decline as their food supply
diminishes and as they are preyed upon by fish and larger plankton.
Nutrients in the warm surface water are exhausted by the earlier pro-
liferation of phytoplankton but accumulate in the cold bottom water
as bacteria dissolve the rain of debris (mainly dead plankton and fecal
pellets) from above. With autumn, the surface water cools and storms
overturn it and mix it with the rich bottom water, giving rise to a
second phytoplankton bloom that is not as vigorous or long-lasting as
that of spring.

As winter follows, the light fails, the waters chill to near freezing,
and life in the plankton is sparse. Nutrients slowly rebuild, and the
cycle awaits the return of spring.

High rates of runoff from the land, both sediments and nutrients,
can alter this seasonal variation, often promoting blooms that cause
"red tide" or phenomenal growth of extremely small creatures. Because each species in the zooplankton and the sessile filter feeders are
geared to catch prey of a certain size, a bloom of an ultrasmall animal
that outreproduces and crowds out all competitors can use up available nutrients, yet provide no sustenance for the food web. This happened in Peconic Bay, Long Island, over several years—to a point
where hardly anything else was present in the plankton. As a consequence, scallops, clams, and fish all but vanished from what was once
bountiful water.

Although small size plays an important role in preserving some
phytoplankton from predation, it is not their only defense. Many are
heavily armored and spiny. The dinoflagellate *Ceratium* is large
enough to be taken by copepods and bigger zooplankters, but is not
heavily grazed even when it is abundant and other food is scarce. At
each end of its triangular body extend three long fishhook-shaped
spines, apparently an unappetizing mouthful. Dinoflagellates and
some blue-green algae are also toxic, and are thus avoided by potential enemies.

For countless thousands of bottom-dwelling animals in the sea, dis-
persing young into the plankton successfully insures their survival
from one generation to another. Consider the **acorn barnacle,**
Balanus balanoides, that bane of boat owners, which develops only on
hard surfaces swept by moving water. Released as a swimming
nauplius about the time of the spring phytoplankton bloom, within a
month it transforms into a *cypris* and begins to look for a permanent

place to settle down. The nauplius and cypris are swimming larval stages, and are as different from each other and the adult as are a caterpillar, chrysalis, and butterfly. During this long course from nauplius to adult, it can be swept considerable distances and can colonize new, remote surfaces.

The **blue mussel,** *Mytilus edulis,* sheds eggs and sperm into the water, and within half a day the fertilized egg develops into the forerunner of a *veliger,* its swimming larval stage. Within a week, the transformation to veliger is complete. Feeding on phytoplankton, it can stay in the plankton for over a month before settling.

Both barnacles and mussels shed eggs and sperm into the open waters simultaneously and en masse. Similarly, many swimming species (herring, for example) and species where the female remains dug in the bottom while the male swims free (the *palalo worm,* for example) also use this technique, which enhances the chances for fertilization.

Vertical Migration

From as far back as the HMS *Challenger* expedition, a curious phenomenon has been observed: Many zooplankton change depth over the course of twenty-four hours. As daylight fades, they swim up, concentrate near the surface for a few hours, then scatter. With the rising sun, they retreat into the depths. They go down to their preferred levels as the sun reaches its zenith. This diurnal migration is caused directly by light intensity; on cloudy days the zooplankton don't descend as deep.

This behavior is widespread among representatives of the animal kingdom in the plankton irrespective of size or phylum. For a small copepod, the length of the trip is astonishing. *Calanus,* no more than three millimeters long, sinks and climbs over ninety meters every day, sinking at a rate of forty-five meters an hour and rising at a rate of fifteen meters an hour.

When the migrating layer is thick and filled with larger plankton—prawns or krill, for example—it can be picked up by sonar. During World War II, the origin of this "deep-scattering layer" was a mystery, but sightings by submersibles verified its biological origin.

The phytoplankton do not migrate; they stay in the surface waters. Why then do zooplankton undergo this enormous expenditure of energy or, put another way, how does it enhance their survival? One theory has it that it offers the plankton their only means of dispersal. The speed and direction of water currents varies over depth, thus, no patch of phytoplankton is apt to be grazed down too long by the same mix of zooplankton and, because zooplankters of different species rise and fall at different rates and seek their own preferred levels, no mix of zooplankters will remain constant for long. Zooplankters eat one another as well as algae, and vertical migration may reduce the chances that one species will overharvest another. Thus, no amount of heavy grazing or heavy predation lasts too long because the protagonists are slowly separated over time, both vertically and horizontally.

You can see the effects of light upon plankton for yourself by placing a diluted portion of your catch in a darkened room. Shine a spotlight on the side of the jar and look down into it. Some plankton converge on the beam like moths to a flame, while others ignore it. One experimenter claims that by adding dilute India ink to a tall jar of mixed plankton and providing light only at the top, the plankton will segregate vertically according to their light-level preferences.

THE FOOD WEB

Like the rhyme about larger fleas having little fleas, the straight-line energy flow theory—from simple plants up through the herbivores on to the carnivores in increasing size, until one comes to the apex predator (big fish, whale, or man)—is a firmly entrenched idea that is only partly true. Each step up, as conventional wisdom has it, requires ten times as much productivity in the preceding step to sustain it.

But the food web in the sea is more complex that that. Many phytoplankton are not eaten at all: they die, sink, and decompose. Only two percent of the sunlight that strikes the sea is converted into plant material, and less than two percent of the energy absorbed from the sun by the phytoplankton is transformed into tissue eaten by animals.

Zooplankton feed not only on phytoplankton but on one another. Some represent a dead-end in the food chain: Jellyfish and combjellies may make a meal for an occasional turtle but seldom for fish. Most, like many of the phytoplankters, simply die, sink, and decompose without boosting a predator in the next rung up the ladder.

Zooplankters have wide differences in longevity and in life cycles. Long-lived zooplankters tend to reproduce steadily and slowly over time with less immediate response to environmental factors than do short-lived ones who can reproduce in enormous bursts, given a favorable environment. Explosive reproduction of small prey favors certain predators over others; a "sticky" feeder like a comb-jelly runs into more food and thus outeats and outbreeds a selective feeder who takes prey one by one.

Because of the uneven dispersal of plankton and fish, often the right food is not at the right place at the right time. An outburst of copepods does little good to fish that are somewhere else. Instead, the copepods may fall victims to a predator like the comb-jelly, who simply ends the chain.

Adult predators that are high up the food pyramid often go through a period of extreme vulnerability in the plankton during their early stages. Fish eggs and fish fry take their chances adrift, along with everything else. The phenomenal variation in the numbers of certain species of fish that reach maturity from a given spawning season has been attributed to the early exposure to enemies. Nature

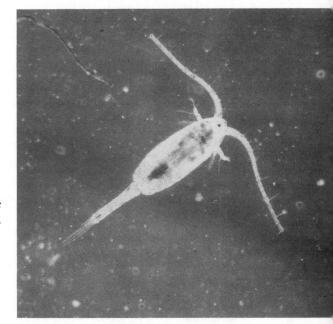

Copepods are a major staple in the diet of young fish.

supplies abundant spawn each year. The problem in growing up is making it past the hordes of waiting opportunists in the sea. Like so many other things in nature, the survival of fish appears to be a matter of chance.

Mankind derives only two to three percent as much food from the sea as from the land, a figure unlikely to improve in the future. The notion of harvesting plankton directly, bypassing intermediate predators, appears logical enough, but in practice is thwarted by their uneven distribution and variable quality. After all, the intermediate predators do us the great service of both finding the prey and selectively eating what is useful.

One possible exception to this may be **Antarctic krill,** *euphausid* shrimp. The Russians and Japanese are fishing for it experimentally. It is the only food source for several species of whale. The blue whale consumes four tons of krill a day when feeding, and takes two decades to reach maturity, gaining sixty tons in the process.

POLLUTION

The effects of pollution on the sea vary with the pollutants. Runoff and sewage bring unusually high concentrations of available nitrogen and phosphorus into coastal waters. Above-normal concentrations of both these nutrients trigger algal blooms, have been implicated in toxic blooms, and in "anoxic events," where the dieback of an algal bloom consumes all the dissolved oxygen in the water, killing the animal life in it.

Runoff and sewage also can contain a wide array of pesticides, aromatic hydrocarbons, industrial compounds, and heavy metals, some of which are innocuous and others highly toxic.

Some of these pollutants are dissolved or remain suspended for long periods of time and are thus carried well out to sea. Others settle rapidly. The life on the bottom of most industrial harbors has been greatly reduced by the poisoning effect of this fallout. Some bottoms are so fouled that the only living things there are polychaete worms. *Capitella capitata,* a polychaete, is a worldwide indicator of polluted bottoms. They colonize barren bottoms through their planktonic larva. Although each adult sheds only a hundred eggs or so, each generation takes only thirty to forty days to grow from egg to breeding adult.

Zooplankton populations diminish in polluted water. Fish fry and eggs are damaged, as are invertebrate larvae. Oyster larvae are especially susceptible to a wide variety of pollutants.

Both chronic and catastrophic oil spills reduce zooplankton; usually larvae are hardest hit, and are either deformed or killed outright.

Only a few of the heavy metals are toxic; indeed some are necessary trace elements for growth. The effects of heavy metals depends on the chemical state of the element. If in a state where it can be readily absorbed, some metals can reach dangerous levels by *biomagnification.* For example, mercury and cadmium can concentrate in the tissues of fish or shellfish who eat plankton that have absorbed them. The metals may not harm the intermediate hosts, but could hold dire consequences for the birds or humans who eat the intermediate hosts.

5

Sandy Bottoms and Temperate Waters

GREAT SPANS OF THE SHORELINES AND SHELVES OF THE CONTInental United States are lined by an enormous boundary of sand that outlines the East Coast below the Gulf of Maine, the Gulf Coast, and the southern tip of the Pacific Coast. From Cape Cod southward, the character of the coastline reflects the forces that brought it into existence and now shapes it. Crushed and worn mineral grains, washed off the land by eons of erosion, fan out over enormous expanses of the continental shelf. Rivers still carry sediments seaward and, where the quantities are substantial, form deltas. In the north, retreating ice left glacial till. The moraines of Cape Cod and Long Island are still being eroded by waves, leaving alluvial cliffs, new beaches, and offshore bars.

This legacy of sand and moving water continually changes the shape of shorelines and bottoms. Direct wave erosion carves long, straight shorelines, while laterally drifting sand creates barrier beaches forward of drowned valleys, embaying the waters between rivers and open ocean. Storms breach these fragile dunes, cutting the ribbons of sand into barrier islands. The shoreline of the United States is dotted with barrier islands from Massachusetts to Padre Island, Texas, that enclose saltwater lagoons and estuaries.

Vast plains of sand lie off the Eastern Seaboard, bound to the seaward edge by the continental slope and the northward flow of the Gulf Stream. From Long Island north, both the coast and the sea bottom were sculpted by glaciation. There, the continental shelf is covered by a series of relatively flat and shallow outwash plains of cobble and sand left by the retreating ice. Between these banks lie basins, troughs, ledges, and hillocks of rock.

From Cape Cod to the Gulf of Mexico, the near-shore bottom is lined with ridges and swales of sand. These ridges are nearly parallel to the shoreline above Cape Hatteras and nearly perpendicular to it below Hatteras.

Submarine canyons transect the broad eastern continental shelf at regular intervals. They were cut by ancient rivers whose valleys have been submerged by the 120-meter (400 feet) rise in sea level that took place between eighteen thousand and three thousand years ago. Along the steeper slopes, the sand may be missing entirely, exposing the underlaying stone, mud, or clay.

Even on flat terrain the bottom may sometimes be bare. If the underlayer is a firm clay, called *marl,* it is often riddled with holes occupied by crab, lobster, or bottom-dwelling fish. The **tilefish,** *Lopholatilus chamaeleonticeps,* has modified hundreds of square kilometers of marl bottom off the mid-Atlantic continental shelf, pocketing the bottom with cavities that have provided cover for an offshore lobster population.

Inshore, the sand is often only a thin veneer over a cobbled or stony bottom, which is periodically exposed by winter storms that move the sand offshore, then replace it the following summer by gentler swells. The sand on the beaches at La Jolla, California, and other places along the western coastline comes and goes seasonally, depending on the size and strength of the Pacific rollers.

Coves along rocky shorelines can accumulate enough sand to form a permanent beach, even though the offshore terrain is mostly cobble, boulders, or ledges and layers of rock. The sand need not necessarily be mineral in origin. Sand Beach on Mt. Desert Island, Maine, contains almost no conventional sand at all, consisting mainly of pulverized seashells and sea-urchin spines.

The difference between boulders, cobble, gravel, sand, mud, silts, and clays is in size, not in composition, although the larger particles invariably originate from rock. Atlantic coastal sand is mainly quartz, often mixed with small amounts of feldspar, hornblende, mica, and

augite. Darker sands that show up along Georgia shores contain il-
menite and rutile, but pure quartz reappears in northeastern and
central Florida. Below Cape Canaveral the sand contains increasing
quantities of calcium carbonate: the remnants of living things like
coral debris, shell fragments, and foraminiferans.

Particles can also precipitate from seawater, forming *oolites:*
globular deposits of concentrically layered calcium carbonate or other
minerals. Calcium carbonate precipitation also cements grains together, transforming loose sediments into a variety of limestones.

On the East Coast, the continental shelf is nearly 160 kilometers
(100 miles) wide off New York and remains relatively broad (100 kilometers, or about 60 miles) as far south as Jacksonville, Florida, where
it narrows rapidly until, at Palm Beach, it virtually disappears.

On the western shores of Florida, the shelf widens to 240 kilometers (150 miles) and gently slopes seaward until it ends at an
abrupt escarpment. Further west, the delta of the Mississippi river
fans out to the shelf's edge. West of the Mississippi, the gently sloping
shelf is rippled by smooth hills and elongated basins. Along the central Texas coast, inshore sandy sediments give way to mud bottoms
offshore. Farther out, the bottom turns to *Globigerina* ooze, created
from the deposits of legions of foraminiferans.

The continental shelf off the West Coast is narrow, and in some
places practically nonexistent. The sandy portions of it lie mainly to
the south. The 100-fathom line (183 meters) lies barely 16 kilometers
(10 miles) off the coast of San Diego, and the bottom is a patchwork
of boulder, cobble, coarse sand, fine sand, and silt.

These waters range from temperate to subtropical. From Cape Cod
to Cape Hatteras, winter surface-water temperatures drop to near-
freezing inshore, but are warmer offshore, as moderated by the flow
of the Gulf Stream. Below Cape Hatteras, winter water temperatures
rarely fall below 10° C (50° F). Off southern California, temperatures
range between 10° C to 24° C (50° F to 75° F), winter to summer.

BURROWERS AND CRAWLERS

Sand relates to particle size—not composition—and particle size de-
termines what life will live within it. It's more than just a question of
whether an animal can successfully burrow into it. Permanent bur-

rowers feed either by drawing in water and filtering out plankton and organic debris, or by swallowing silts and muds, extracting nutritive needs from the organic matter in them. Watery silts and muds work to the disadvantage of suspension feeders by clogging their intakes. Sediment ingesters, those animals that feed on the organic matter and small living forms between mineral grains, thrive where the suspension feeders fail, but seldom occupy coarser sand because its nutrient value is low and its grains are too big to process.

Coarse sand is home to both permanent and temporary burrowers. The **surf clam,** *Spisula solidissima,* the largest bivalve on the East Coast, digs in here. It is the basis of an important fishery. Usually the surf clam lies totally buried from view except for a telltale depression in the sand where its siphon protrudes, inhaling and exhaling water for food and oxygen. Because coarse sand can shift with currents and ground swell, the clam is sometimes partially exposed, often remaining that way long enough to take on a green patina of algae above the sand line.

If you pull a specimen out and lay it flat on the sand underwater, it will begin to reburrow immediately. The clam will open its shells, extend its large muscular foot, and push down into the sand, simulta-

Surf clams are harvested by dredging. Jets of water carry the sand away from the steel sled.

neously rolling itself into a vertical position, hinged-end up. With a series of foot thrusts, it begins to dig in. The clam jets away the sand at the base of its foot by opening its shells, letting in water, then sharply snapping the shells shut, sending a strong pulse of water downward. Eight to ten rapid contractions and the job is done.

Cold water significantly slows the clam's maneuverability. Winter gales often uproot and dump thousands on northern beaches where, exposed to the air, they quickly freeze and die.

The distribution of various kinds of clams depends on wave energy. The West Coast **pismo clam,** *Tivela stultorum,* will live only in surf. Dug up and transplanted to quiet water, it quickly perishes. The **razor clams,** *Ensis* and other genera, prefer calmer waters. A razor clam, though it is thin-shelled, can rebury itself in less than seven seconds.

For sheer burrowing depth, nothing equals the West Coast **geoduck,** *Panope generosa,* which lives in soft mud at a depth of 1.3 meters (4 feet). The shell may be twenty centimeters (eight inches) long. The geoduck (pronounced "gooeyduck") can weigh as much as 5.5 kilos (12 pounds). Neither as large nor going as deep, but still a challenge to extract, the **gaper,** *Tresus nuttalli,* lives from one-half to one meter below the bottom. It rarely weighs more than 1.8 kilos (4 pounds). Both the geoduck and the gaper are slow diggers, although their quickly retreating siphons give the illusion that they can withdraw rapidly.

Bivalves

The bivalves of sandy and muddy bottoms are legion, both in number and species. Hunted by man as food, every coastal state has rules and regulations concerning where, when, and how commercial species may be harvested. The inshore waters of the East Coast were once famous for their abundant yields: scallops from Peconic Bay, quahogs from a multitude of Long Island coastal embayments, oysters from Chesapeake Bay, and so on down the seaboard. But human encroachment along the coast has inexorably led to overharvested and polluted bays and estuaries, exhausting formerly productive grounds or rendering their shellfish unfit for consumption.

Bivalves are filter feeders and can capture particles the size of bacteria, as well as viruses attached to larger particles. They can extract carriers of human disease from impure water without harm to themselves and pass them on to the unfortunate persons who happen to eat them. They also have the uncanny ability to extract chemicals, heavy metals, and oil from waters tainted with industrial waste. Long before state regulators closed parts of Raritan Bay in New Jersey to shellfishing, the industry had suffered because customers claimed the clams tasted of the coal oil (kerosene) that leaked from nearby refining and transfer operations.

SCALLOP

Most adult bivalves make for dull additions to a saltwater aquarium. The exception is the scallop. The scallop family has over three hundred members worldwide. They do not burrow, but lie on top of the sand. Colorful and active, their mantle margins are lined with iridescent blue eyes. The hinged valve is held by one large *adductor* muscle (instead of the usual two small ones in other bivalves), whose quick contractions can propel the creature well off the bottom when necessity dictates, as it so often does when the probing arm of a sea star demands a fast response.

The bay scallop, *Aequipecten irradians,* has thirty to forty bright blue eyes along its mantle margin. It responds quickly to sudden changes in light.

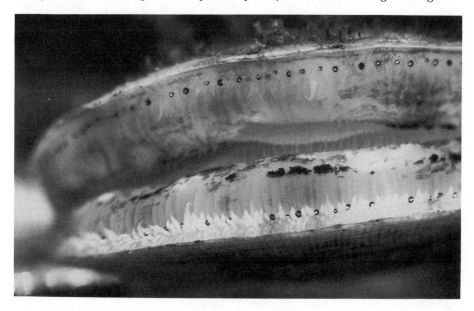

Keeping a sea star and a scallop in the same tank for any length of time will eventually result in the scallop's demise. A home aquarium is too small to accommodate both these ancient enemies. The same is also true for a long list of other combinations of creatures. If you want to observe a number of bottom dwellers, the answer is two aquaria, and even then you may have to partition off sections for vulnerable species. Some, like the sand dollar, have very limited means of evading trouble.

Sand Dollar

The **sand dollar,** a flattened disc-shaped relative of the sea star and sea urchin, litters wide expanses of bottom where the sand is coarse enough to burrow in quickly. *Echinarachnius parma* is the most common one in the Northeast, the **keyhole urchin** *Mellita quinquiesperforata* the most common in the Southeast, and *Dendraster excentricus* the most common on the West Coast. They are a favorite food of haddock, cod, flounder, and California sheepshead, whose stomachs are often crammed full with their crushed remains.

The connection between the sand dollar and the sea star can be seen when you look at the sand dollar's bleached and dried outer skeleton (called a test); the five-rayed pattern of pore openings prominently stands out. Aside from the ray display, like other echinoderms it possesses movable spines, movable jawed pincers, and tube feet.

Sand dollars and key urchin feed in and on top of the sand, working over the grains for algae and organic debris. If something edible lands on their upper surface, they can detect, capture, hold, and transfer it to their mouth. Considering their shape and the position of their mouth (in the center of the underside of the disc) this takes some doing.

On the top surface, the spines of the sand dollar are clubby and covered with cilia. Mucus is exuded at the base of the spines and flows into the nearest food groove. Any small particle ensnared in it inexorably moves down the groove, over the edge of the disc, then into one of the five main food grooves running to the mouth. A large particle is grabbed by a tube foot and passed to a groove.

On the underside, the thin, stiff spines encircle the morsel teepee-fashion, holding it fast until the jawed pinchers can grab it. Then the tube feet get hold of it and transfer it to a food groove.

The sand dollar, *Echinarachnius parma,* often litters open sand bottom, and is an easy meal for cod, flounder, and haddock.

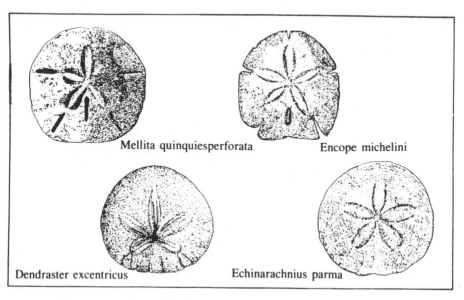

Mellita quinquiesperforata   Encope michelini

Dendraster excentricus   Echinarachnius parma

Sand dollars *(Illustration by Michele Cox)*

You can follow this process at home. Put a sand dollar into a shallow dish of seawater. With any common food dye, color some bits of fish or finely chopped fresh clams or plankton. Place a few particles on the upper side of the sand dollar and watch their movements. Entrapping and moving food via a mucous stream is done by some bivalve mollusks, gastropods, polychaetes, bryozoa, brachiopods, and a number of other bottom-dwelling suspension feeders.

The reaction of the sand dollar to the approach of an archenemy, the **common sea star,** *Asterias forbesi,* shows that it can chemically sense a predator. Plunk a large sea star down in the midst of a bed of exposed sand dollars. Those downstream from the current flow will immediately burrow under the sand. As the sea star moves, those in its path quickly go below. Only sea stars that feed on sand dollars have this effect. For some sand dollars, as well as clams, burrowing doesn't always help; there are species of sea stars that burrow to find their prey.

Sea Star

Sea stars are well-known enemies of the clam. Tube feet in the body grooves on a sea star's undersides end in sucking disks that, powered hydraulically, can clamp around a clam and exert a pull of over a hundred pounds. Not all clams are easy victims. The **giant Atlantic cockle,** *Dinocardium robustum,* and the **basket cockle,** *Clinocardium nuttallii,* common in Puget Sound, can escape the sea star by leaping. They have a long, curved, heavily muscled foot that they can straighten out so violently that they propel themselves a half-dozen body lengths off the sand before the sea star can gain a hold.

In the sea, life attaches to nearly every smooth or hard surface it can find. Algae hydroids, bryozoans, and the like encrust animate and inanimate objects with equal vigor. How then does a sea star keep growths off itself? Some sea stars burrow through the abrasive sand, scouring off anything that settles on them. Others that do not burrow have a mechanical way of keeping small organisms from attaching to them. Projecting from their top surface, little movable pincers grab and crush any offending objects that settle there. You can test this for yourself (harmlessly) by inverting a common sea star on to the back of your hand. As you remove it, you will find it has grabbed the hairs (but not so hard that you can't easily pull it away).

It's curtains for this mussel once it's in the grip of the sea star *Asterias*.

Whelk

Clams have other enemies: everything from large skates and rays who uncover them by plowing through the bottom, to the slower, more deliberate **horseshoe crab,** *Limulus polyphemus,* who does much the same thing when looking for small bivalves. All crush their victims.

The **knobbed whelk,** *Busycon caricam,* the **channelled whelk,** *B. canaliculatum* (the largest whelks on the East Coast), and the **moon snail,** *Lunatia heros,* seize their prey with their muscular foot, rasp a neatly beveled hole through the shell, then insert a feeding tube that sucks out the clam's tissues. Another gastropod, the **oyster drill,** *Urosalpinx cinerea,* much smaller and more numerous than *Busycon,* preys upon the exposed and sessile oyster in much the same way.

You will occasionally come across a female whelk laying eggs. The egg case emerges as a chain of parchmentlike discs, as many as one hundred to a string, that itself may extend to a meter in length. Each disc holds several dozen miniatures of the adult.

A channeled whelk laying eggs. Normally the egg cases
are buried underwater in the sand.

The moon snail lays its eggs encased in sand. The sand collar, a
thin, circular, raised sheet of glued sand grains and eggs, is nearly
fifteen centimeters (six inches) in diameter. A piece of that collar
placed in an aerated dish of seawater will quickly develop into hundreds of miniature moon snails that can easily be seen with a magni-
fying glass. There are so many the sand seems to seethe as they begin
to hatch.

The small hermit crab, *Pagurus longicarpus,* inhabits empty periwinkles and small moon-snail shells.

Hermit Crab

Empty moon snail and whelk shells often become the homes of large **hermit crabs,** particularly species of *Pagurus.* Hermit crabs have a soft, unshelled abdomen, too vulnerable to withstand attack, that they protect by backing into the right-sized shell. Suitably fitted out, the hermit's large right-hand claw, when withdrawn, will completely block entry to the shell. Large or small hermit crabs make marvelous additions to a home saltwater aquarium. Be sure to supply them with a choice of shells. As they grow they must continually search for something the next size up. Apparently they can distinguish between a shell they have already investigated and new prospects. They will look over the old ones again and again, but not with the care that a new one engenders.

In your search for hermit crabs, you are sure to come across some with a pink, fuzzy growth over their shells. This is the hydromedusan *Hydractinia,* a remarkable little cnidarian that does quite well in a salt-water aquarium. Under a low-power microscope you will discover it has five different upright structures (called *zooids*): one sensitive to touch, one for defense, two that feed, and one that generates eggs, which settle down nearby to produce new *zooids.* The zooids are inter-connected at their bases by *stolons,* which exchange food and pene-trate the shell, firmly anchoring the animal.

The mutual advantages of this relationship are hard to decipher. The hermit crab will occasionally wipe the shell with its claw and carry off a snack of plankton caught by the zooids. In turn, the crab is constantly turning over and stirring things up so the hydromedusan probably derives many a meal from the fallout. As the hermit crab grows, however, it will seek another shell and abandon his hitchhik-ing companions who, unless quickly taken up by another hermit, will expire.

Anemones often hitch a ride on the shell homes of large hermit crabs. In southern waters you will commonly find the pink- and brown-striped **cloak anemone,** *Calliactus tricolor,* on board. This is no small load, either; the shell may house up to twenty small anemones on it or just one big one, often so large its pedal disc covers most of the shell.

In the hermit crab world, having an anemone on your back must be something of a status symbol. If you put several hermits together in an aquarium, some having an anemone and some not, you will soon see the have-nots looking over the shells of the haves. By tap-ping the shell of the have, then grasping the anemone (which imme-diately lets go of its hold on the shell), the have-not transfers it to its own shell. The former have makes no fuss about this redistribution of wealth, but soon starts looking over the crowd for another traveling companion.

Anemone

A few burrowing anemones make their homes in the sand. In the south, the **sea onion,** *Paranthus rapiformis,* and a related Pacific species, *Pachycerianthus,* settle in coarse sand and can survive being uprooted by surf. In quieter, siltier waters, *Cerianthus* and *Ceriantheopsis* species placidly lie buried up to their tentacles, which they extend outward radially and flat on the bottom. They will withdraw them with lightning speed if touched by anything threatening.

Quieter sands and silts host multitudes. You are apt to see, lying on the surface of the bottom, the **sea pansy,** *Renilla,* a drab, flattened-heart-shaped cnidarian whose surface is covered with clusters of little polyps, giving it the appearance of a flower bed. It is anchored in by a body part that projects down into the sand. These are close relatives of the *sea pens* and *sea feathers,* upright deep-water forms that are substantially gaudier and more colorful. Sea pansies are luminescent, glowing a faint fiery green in the dark. If you look closely you will see the glow emanates from tiny brightly glowing points on their bodies. They do well in aquariums.

Sea Cucumber

Fine sand and silt hosts the sea cucumber, the least likely member of its five-sided, star-shaped, echinoid clan. Basically a tough skinned bag open at each end, with one end tentacled, it either burrows through the bottom, extracting its needs, or buries itself with tentacles waving, catching its meals from the passing plankton.

What makes a sea cucumber an echinoderm? Without plunging into the similarities in architecture in too much detail, the most prominent commonality is tube feet that are either all over the body or concentrated along one side. (Like all generalities, this one isn't universally true. Some sea cucumbers have no tube feet.) The sea cucumber's tube feet have the same function they do for the sea star or the brittle star: locomotion.

Another feature (again, variable among species) is the calcareous plates, or *ossicles,* that lie just beneath the skin. The **rough-skinned thyone,** *Thyone scabra,* bristles with them: flattened platelets riddled with holes that have a raised projection in the middle of each one. Other sea cucumbers, species of *Cucumaria,* not only contain rectangular plates but also others more elongated, shaped like errant coat hangers.

The shapes of these plates can be a help in telling species apart. In *Synapta,* the plates are anchor-shaped and microscopic in size. Many species of *Synapta* are translucent, and if you can recover one whole, you can see its internal machinery without harming it. Most are thin, eight millimeters in diameter and an average fifteen centimeters in length.

Although the branched feeding tentacles in many sea cucumbers are dull olive or brown, some are quite gaudy and present a spectacular sight underwater—bright patches of pink, red, yellow, or purple—against an otherwise drab background.

Worms

Sandy bottoms in quiet water ten to twenty meters deep can be littered with empty sand dollar tests. The remains of the departed provide firm footing for hydroids, bryozoans, serpulid worms, and other opportunists who otherwise would be hard-pressed to find a place to get a start in life.

In southeastern and Gulf waters, you may find the inside of the test holds an echiurid worm, *Lissomyema mellita,* that has grown too big to get out. *Lissomyema* belongs to a small phylum of spoon worms, *Echiura,* who have only a handful of relatives along our coasts: the **landlord worm,** *Thalassema hartmani* in the East, and the **innkeeper,** *Urechis caupo,* off California. These dig U-shaped burrows that host other animals: clams, crabs, scaleworms, and, in California, a goby.

Sharing a home with its builder is common in the sea. The **lugworm,** *Arenicola cristata,* and the **parchment worm,** *Chaetopterus variopedatus,* two very common mud burrowers, invariably share their digs with small pea crabs.

The parchment worm is common on both coasts. Its U-tube openings are spaced an average of twenty-five centimeters apart: characteristic hollow white-lined tubes that protrude above the bottom. Although not an easy creature to keep in an aquarium, transferring it to a glass tube will allow you a few days to see how it pumps water through its burrow in a sort of three-part piston arrangement rather than the more common pulsing action of other bristle worms. When touched, it glows with a soft fire blue.

If you swim over a smooth silty bottom, you will be struck by the number of holes and mounds that dot the seascape. They are mainly the homes of worms, although a few crabs also burrow in. Beneath, there are legions dug in without a trace on the surface. Polychaetes are among the more prominent. Some burrow beneath the surface, but many build themselves a tube dwelling. These either create a smooth-walled hole from an excreted cement, or use it to construct a tube of sand grains or whatever other material is available.

The **plumed worm,** *Diopatra cuprea,* casts a tube that extends seven to ten centimeters above the sand and is decorated by fragments of shell, seaweed, and, as likely as not, various forms of life: hydroids, serpulid worms, bryozoans, and, occasionally, barnacles.

Other worms simply crawl over the surface, half-buried. The **sea mouse,** *Aphrodita hastata,* an oval-shaped blob covered with hairs, makes its way by undulating along, leaving a characteristic trail behind it.

The number of holes, mounds, fecal casts, depressions, and trails made by bottom life is ample evidence of the diversity and sheer quantity on and in the sea bottom. To find out what is there means digging and sieving: a shovel and screen will do in shallow and deeper water. A small galvanized pail for a scoop and a large plastic bucket with a rope to pull it up will hold enough to keep you busy sieving, picking, and identifying. Your screens may range anywhere from a frame of hardware cloth to a relatively fine kitchen flour sifter, depending on the coarseness of the sediments.

Crabs

The **rock crab,** *Cancer irroratus,* in the Northeast, and the **calico crab,** *Hepatus epheliticus,* to the South, scavenge over open sand. They burrow in with just their eyes showing during the day as they wait for darkness or an easy meal to pass by. On the Pacific Coast, the **Dungeness crab,** *Cancer magister,* found normally in deep sandy bottoms, comes inshore to molt and breed. A commercially valuable species, Dungeness catches gyrate wildly from year to year. However, what appears to be a permanent decline has taken place along the Central California coast. Reasons range from sewage and industrial waste, to microbal fouling of the eggs, to introduced fish preying on the larvae, to what is most likely, parasitism by a nemertean worm on the eggs. Biologists estimate that over fifty percent of the eggs of the Central California population are destroyed by this worm each year.

Brittle stars can move
quickly, snatching worms
and other prey, but are, in
turn, preyed upon by fishes.

The *brachyurans,* true crabs, have adapted to hundreds of bottom
niches in the sea. The **blue crab,** *Callinectes sapidus,* is a swimmer as
well as a fast crawler, and is so well suited to so many bottom types
that it has become prolific along the whole East and Gulf coasts. Agile
and rapacious, with absolutely no sense of humor, a big male *jimmie*
or a female *sook* are quick to snatch bait, or your hand if you are not
careful. In the North, when the waters grow cold, they burrow in
mud and lie dormant until spring, but are still prey to the dredges of
crabbers. In the South, they are active year-round.

Brachyura means "short-tailed." True crabs are either rounded or
squarish. Their abdomen has been reduced to a small flap tucked
beneath the body. They have five pairs of legs, and the first pair is
equipped with pincers. Most are scavengers, although a few, the pea
crabs, can snatch plankton from the water. Large swimming crabs can
also grab a passing live meal now and then.

On muddy and sandy silt bottoms you are apt to meet the sharp-
nosed spider crabs. They have the curious habit of decorating themselves with seaweed, sponges, bryozoans, tunicates, and any other ses-

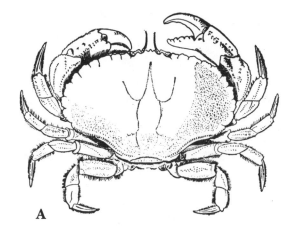

A, The rock crab, *Cancer Irroratus*
B, The Jonah Crab, *C. Borealis*

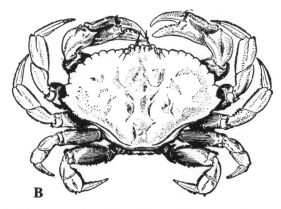

These two look-alikes are often found together in northeastern waters. The Jonah is the larger; its claws and shell are rougher, the marginal teeth along its carapace are rougher, and its color is more purple. *(Illustration from A. F. Arnold)*

sile forms (including bits of plastic) that are handy. This disguise isn't a show of intelligence, but simply a matter of instinct. Put one in an aquarium and let him go hungry. He will eat everything edible on his back. Now feed him and scatter in a few pieces of bright plastic. The crab will seek them out and put them on his back, even though they will make him more conspicuous than before.

Because they are so active, adding a crab to your aquarium seems like a good idea. And so it is; a crab will certainly keep it clean of

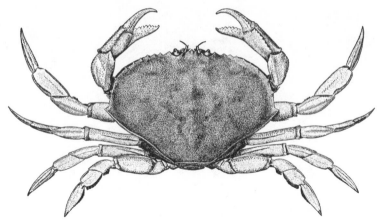

Cancer magister, the common crab of the Pacific coast; male. *(Illustration from A. F. Arnold)*

debris. However, most are diggers and you will find the bottom rapidly rearranged to suit *their* needs, not those of the aquarium. If you put a swimmer, say a small blue crab, in with fish, the crab will hunt them relentlessly (unless, of course, the fish are big enough to turn the tables and hunt the crab). Small fish will get no rest, night or day, and will become so skittish they will flee at every moving shadow. You will lose some to the crab and some to the floor, because in their panic to escape they will leap clear of the water, often right over the edge of the tank.

SWIMMERS

At first sight, underwater stretches of sand seem deserted. But if you can arrange to be slowly towed over a sand bottom, you will discover life there to be active and prolific.

On open sandy sea bottom, very little protective cover exists, and whatever *is* there is put to use. Juvenile fish hide in discarded tin cans, under shells, and will even curl up within a moon snail's sand-collar egg case. Young **red hake,** *Urophycis chuss,* can often be caught by plunking a mesh-bag net over a sand collar or surf-clam shell, then shooing the fish into the bag.

Every exposed stone harbors an anemone or stalked hydroid. If the

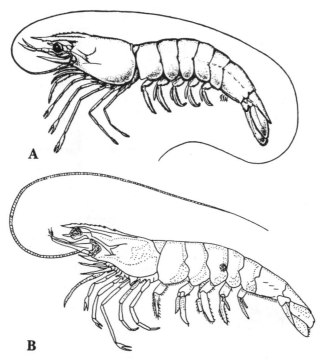

A

B

a, the white shrimp, *Penaeus setiferus;* and *b,* the pink shrimp, *P. duorarum,* are commercially sought in the southeastern United States. The white shrimp are plentiful from Cape Hatteras to northern Florida; the pink shrimp in Florida Bay, the west coast of Florida, and the Gulf of Mexico. *(Illustration courtesy of Fish and Wildlife Service)*

water is shallow and clear, sea grasses grow and are a haven for countless small fry. In bays and estuaries especially, these grasses harbor the **grass shrimp,** *Palaemonetes vulgaris,* a favorite food of flounder and fluke, who stalk it by slowly undulating over the sand, arching their midriffs, and striking with the speed of a snake.

Flounder and fluke often lie buried in the sand with just their eyes protruding. So do skates and rays. Their camouflage is so complete that you may inadvertently find yourself nearly on top of one before it explodes from the sand beneath you in panic.

Unlike the grass shrimp, the commercial shrimp of the Southeast and Gulf coasts—the **pink shrimp,** *Penaeus duorarum,* the **brown shrimp,** *P. azecus,* and the **white shrimp,** *P. setiferus*—swim in vast schools. The white shrimp is the mainstay of the fishery off the Louisiana and Mississippi coasts.

For these shrimp to reach maturity, water currents and salinity must be just right, for their lives are short and not amenable to sudden changes in plan. The female spawns in open ocean but relatively close to shore, near the mouth of an estuary. The water must be very salty, thirty-five parts per thousand. The fertilized eggs sink to the bottom, and in twenty-four hours hatch into a *nauplius*, a tiny initial stage that looks nothing like the adult. Within another thirty-six hours, this creature has gone through four molts and become a glassy *protozoea* (not *protozoa*) which continues to metamorphose into a *mysis*, then on to something that begins to look like its parents. Each molt produces a larger and more structurally complex animal than the last.

The entire process takes three weeks. During this time, the larva drifts helplessly toward the estuary, an area of lower salinity. By the time it can crawl and make headway, it is just under a centimeter long. It seeks out the shallow brackish waters of the bayous and creeks that will be its nursery for the next four to eight weeks. It then moves toward deeper water: the rivers and bays that lead to the ocean. By summer's end it is thirteen centimeters long and heads for open sea. By spring, it has reached maturity, and spawns. Few live to see a second season.

What happens to the eggs not carried shoreward? They perish. This odd confluence of breeding, salinity, and current flow either comes together successfully, or the population crashes, to be rebuilt again by the few survivors that escape the consequences of natural calamity.

North of Cape Hatteras, cold weather and cold water lure **Atlantic cod,** *Gadus morhua,* and **pollock,** *Pollachius virens,* southward in winter. Cod are voracious and egalitarian feeders, eating almost everything that comes their way. On Georges Bank, a favorite feeding ground, the bottom is covered with small cobbles, and these stones are often found in the cod's stomach. The old salt's story is that cod take on ballast before a storm, but the likely explanation is that they swallow the stones with the anemones they uncover while rooting along the bottom.

In winter, *hakes,* deep-water fish, move inshore, as do **winter flounder,** *Pseudopleuronectes americanus.* Other fish migrate offshore: **summer flounder,** *Paralichthys dentatus,* and **black sea bass,** *Centropristis striata,* among them. The northern wrasses, **tautog,** *Tautoga onitis,* and **cunner,** *Tautogolabrus adspersus,* remain around their usual haunts— jetties, rock outcrops, and wreckage.

Atlantic cod, *Gadus morhua*

Bluefish, *Pomatomus saltatrix*

Pollock, *Pollachius virens*

Striped bass, *Morone saxatilis*

Squirrel hake, *Urophycis chuss*

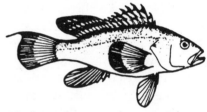

Black sea bass, *Centropristis striata*

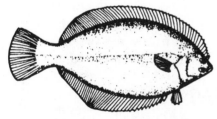

Winter Flounder, *Pseudopleuronectes americanus*

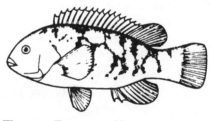

Tautog, *Tautoga onitis*

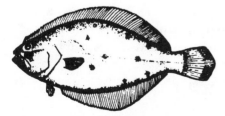

Summer flounder, *Paralichthys dentatus*

Cunner, *Tautogolabrus adspersus*

(Illustrations courtesy of National Marine Fisheries Service)

As the spring waters warm up, **Atlantic mackerel,** *Scomber scombrus,* migrate in from offshore, suddenly appearing, then vanishing seaward again only to reappear, usually north of their previous positions. **Bluefish,** *Pomatomus saltatrix,* follow, going north at a slower pace than the mackerel. Because blues occur over such a vast range, worldwide in fact, they were once thought to migrate north-south along our coasts, but deep-water research trawls now suggest an inshore-offshore pattern is more likely. Off Florida, they vanish in summer, just where is yet unknown.

Weakfish, *Cynoscion regalis,* and **striped bass,** *Morone saxatilus,* migrate northward in summer, hugging the coastline. Neither are ever found very far from shore, and most are caught in surf.

Stripers occur inshore from the St. Lawrence River in Canada to the St. Johns River in Florida, from western Florida to Louisiana, and from Central California to Oregon. Their major breeding grounds are the San Joaquin-Sacramento Delta on the Pacific Coast, and the Chesapeake Bay and Hudson River on the East Coast.

In the Chesapeake, stripers winter in the deep waters of the bay. The immature fish leave in early spring, traveling northward. Adults move upriver to spawn. That done, they go downriver, some staying near the mouth of the bay and others migrating up the coast. Where they go upcoast is hard to tell. Some travel fast and far to Nova Scotia. In fall, they head home to where their journeys began.

Game and food fish get most of the attention of wildlife biologists, sportsmen, and commercial fishermen. Most are relatively large open-water swimmers and, as adults, not suited to small-aquarium living. The naturalist's tank is too confining for fish constantly on the move. There are, however, a great many small fish, some swimmers, others bottom-dwellers, who adapt well to aquarium life and have something to teach their captors.

Juveniles of any species will do for a while. Raising a skate or shark from an embryo found in its egg case will give you an opportunity to see an oceanic wanderer mature, but it won't take long before it reaches a size where you will want to return it to the sea.

Any number of species of anchovy, smelt, killifish, mummichogs, gobies, and sticklebacks can be netted in shallow water and do well in an aquarium. A few sweeps over eelgrass can bring in pipefish and sea horses. Poking around in deeper water, you can often catch a member of the toadfish family, who will grunt, croak, or whistle on occasion. Be careful of its venomous dorsal spine, however.

On the Pacific Coast, the **spotted batfish,** *Zalieutes elater,* can often be spotted "walking" across the bottom on its modified pelvic fins. It has a dorsal fin that functions as a fishing lure, and dangles a fleshy bait in much the manner of the frogfishes and the goosefishes.

If you make the acquaintance of a commercial fisherman who trawls in deep water (one hundred meters or deeper), you may be able to get specimens (alas, usually dead) of deep-sea creatures. The family names of these denizens conjure up their odd forms: stargazers, gulpers, Dorys, chimeras, lanternfish, spookfishes, hatchetfishes, loosejaws, slickheads, pearleyes, viperfishes, snaggletooths, dragonfishes, and others who look every bit as odd as their names suggest.

You can help unravel the life histories of fish—how fast they grow, how far they travel, how long they live—by tagging them. Catch them, record simple but vital data, put on tags, and release them. When recaptured, and more data is recorded, a picture will emerge about their lives.

The American Littoral Society will supply tagging kits for a nominal charge, with instructions on how to tag. When a fish is captured, both the person who released it and the person who recaught it are notified of its travels and increase in size. All the information the Society garners is computerized and fed into a data bank at the National Marine Fisheries Service at Woods Hole, Massachusetts. It is also published in the Society's periodical, "Underwater Naturalist."

6

Rock Bottoms and Cold Waters

Rock bottoms support a very different cast of characters than does sand. Clingers replace burrowers. Surface browsers abound. Inshore, kelps and seaweeds that can grip the rock in the roughest seas provide hiding places and homes for many who would be easy pickings if left exposed.

Much of our rockbound coasts are washed by oxygen-rich, nutrient laden cold water that is, itself, a rich grazing ground for plankton feeders. The chilly waters from the North do not harbor the diversity of species found in the tropics, but what they lack in kind they more than make up in quantity. Nowhere else is animal life more abundant.

THE PACIFIC COAST

The Pacific continental shelf is much narrower than the East Coast shelf; no more than sixteen to thirty-two kilometers (ten to twenty miles) wide north of San Francisco and half that southward. Bottom terrain shallow enough for diving rarely extends more than 3 kilometers from shore.

A few wide shoal areas do extend farther offshore here and there: off San Diego, Los Angeles, San Francisco, Crescent City, and Point St. George on the California coast. Off Southern California several groups of islands—Santa Cruz, Santa Catalina, Santa Barbara, Los Coronados, and smaller islands nearby—can all be visited by divers from boats.

Along much of the coast, winter brings rough seas and turbid water. From San Francisco northward, winter water temperatures average 8° to 10° C (46° to 50° F). Summer water temperatures are not much higher, 11° to 14° C, but calmer seas prevail through autumn and water clarity improves significantly.

Tidal ranges along the Pacific Coast average 1.7 meters from San Diego to Point Arena, then increase to 3.3 meters in Elliot Bay, Puget Sound. Like the East Coast, tides occur twice daily, but unlike the East Coast, the tides are mixed: unequal in height. Thus, one tidal cycle registers the day's extremes for both high and low water; the other cycle is neither as high nor as low.

Kelp

The most striking biotic feature of the cold rock-bottom regime is the algae: kelps on the West Coast and seaweeds on the East Coast. The word *kelp* has changed meaning with time. It originally referred to the ashes of large seaweed burned to obtain iodine, but now means the seaweeds themselves and, in particular, the large broad-bladed laminarians. These are brown algae. Forty or so species grow along the California, Oregon, and Washington coasts. The forests of these seas are made up of three genera: *Macrocystis,* the **giant kelp;** *Pelagophycus,* the **elk kelp;** and *Nereocystis,* the **bull kelp.** Great stands of these three exist in waters well over thirty meters deep.

Kelp harvesting is big business in California. The preponderance of the crop is giant kelp. From it, *algin,* a colloidal agent, is extracted and used for everything from a smoothing agent for ice cream to a slipping agent for oil-well-drilling mud.

Giant kelp grow long *stipes* that are held aloft by gas bladders. From each gas bladder a single large blade arises. Two or more stipes may originate from a *holdfast,* a tangled rootlike system that grips the rock bottom. A mature *Macrocystis* can have a holdfast the size of a bushel basket anchoring several stipes and a bladder-blade system that rises

A beached elk kelp. Note the bladder that holds the stipes aloft and the large holdfast. *(Photo by P. C. Parker)*

sixty meters to the surface. It grows at the rate of 0.3 meter (1 foot) a day—faster growth than bamboo.

Bull kelp, *Nereocystis,* is the most conspicuous of the western seaweeds because it occurs in large beds close to shore. Large strands of the bladders and blades lie on the sea surface. If free-diving, be careful around nearshore kelp beds. Watch for heavy wave action or fast tidal currents. Surge or current may sweep you unexpectedly into a cluster of fronds, from which escape may be difficult.

Deep-water kelp exploration requires scuba gear and a full wet suit. Enter by descending to the bottom at the edge of the underwater forest; if you can, exit by the same route. You may find yourself getting tangled in fronds; simply back up and unhook yourself one frond at a time. Remember that changes in surf intensity happen quickly along the north Pacific Coast. Take no chances along coastline with a bad reputation for sudden, violent surges. Get local advice before diving. Diving near rock outcrops or cliffs where wave action is

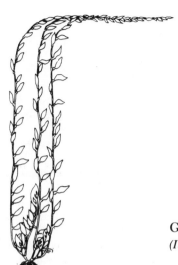

Giant kelp, *Macrocystis pyrifera*
(*Illustration courtesy of "U.S. Fish and Wildlife Service"*)

strong presents another potential hazard. Big logs drifting southward from British Columbia often hug the surf line. Chances of getting caught between a rock and a big piece of moving timber may appear slim, but the risk is there and has proven fatal.

The life cycle of a simple algal plant is complex. Many plants alternate between sexual and asexual generations, thus you can have three individual plants: a male, a female, and an asexual. The neuter plant, called a *sporophyte,* releases sexless zoospores that settle down and form sexual plants. When these mature, they release *gametes* (sexual products akin to ova and sperm in animals) that fuse, settle out, and grow into a sporophyte, closing the cycle. **Sea lettuce,** *Ulva,* reproduces this way. There is no way of telling when you pick up this simple green sheetlike alga whether it is male, female, or a sporophyte.

In the bull kelp, the big plants are always sporophytes. They release zoospores that settle to the bottom and form microscopic sexual plants. The male releases sperm into the water, which must then find a fertile plant bearing eggs that stay put. Arrangements like this would seem to pose a survival problem, but one single bull kelp plant can produce nearly four trillion zoospores a season.

Off Southern California, three species of elk kelp live in deep water. One of these, *Pelagophycus giganticus,* grows a single stipe from

A TASTE OF THE SEA

Almost all seaweeds are edible and downright good for you, being chock full of algin, vitamins, minerals, and trace elements you didn't know existed. The trick is to make them palatable. Just munching on a piece of wrack won't do; it tastes like a salty rubber band.

Dulse, *Rhodymenia palmata,* common along the Northeast Coast, when air-dried for a week can be chewed like tobacco, and also like chewing tobacco is an acquired taste.

Kelp, *Macrocystis,* sun-dried and powdered can be used as a tasty substitute for salt. Oarweed, *Laminaria,* sun-dried (first remove the stipe), can be chopped fine, boiled, and added to soup to thicken it, and can be used as a seasoning (it is sold under the name *Kombu* in gourmet shops).

Some wracks are said to be good steamed, especially if then sauteed in oil and soy sauce. Traditionally, wrack has been used in clambakes to impart flavor to clams and corn steamed in it.

Irish Moss, *Chrondus crispus,* found in tidepools from Labrador to North Carolina, can be used to thicken soups, stews, and to make gelatin desserts. To find out how, look up E. Gibbons *Stalking the Blue-Eyed Scallop,* 1964, David McKay Co., NY.

a holdfast to a large gas bladder from which six to twenty blades arise along a stalk, which gives it an antlerlike appearance. Individual blades can be ten meters long. It thrives around Santa Catalina Island in eighteen to thirty-six meters of water where bottom temperatures range between 11° to 17° C.

Recent warm-water intrusion and rough winter storms have destroyed much of the existing elk kelp beds off Southern California. Attempts to transplant elk kelp into barren areas have run into a curious snag. The adult plant will survive a round trip from a thirty-meter bottom to the surface and back down again, and, if properly reattached, will grow and prosper. But juveniles will not. On the return trip down, their gas bladders collapse from the increased pressure (at thirty meters, the pressure is nearly four times that of the surface). Few survive the trip, and those that do will not grow.

Like all plants, kelp need light. Dense kelp cover in the upper waters forms a canopy over the environment below that cuts off ninety-nine percent of the light that would otherwise reach the bottom. Light levels in some kelp beds are below that needed to sustain growth by photosynthesis, yet replacements do grow. This puzzle remains to be explained.

The protection and cover afforded by kelp fronds creates a snug harbor for over one hundred species of fish. Fish that normally stay close to the bottom in open areas forage through the upper labyrinths of the forest of fronds. The yellow-and-orange **señorita,** *Oxyjulis californica,* and the **sheepshead,** *Semicossyphus pulcher,* hunt through the upper canopy. The sheepshead can live to be over fifty years old, but the decline of kelp forests and spearfishing has substantially reduced its numbers. Like many other wrasses, it starts life as a female, changing to a male as it grows older.

On the bottom, one of the larger sculpins, the **cabezon,** *Scorpaenichthys marmoratus,* has also suffered from heavy predation by divers. Its flesh is quite tasty, but its eggs are poisonous enough to make humans violently ill.

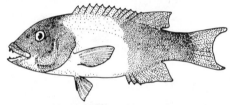

California sheepshead
(Illustration courtesy of California Department of Fish and Game)

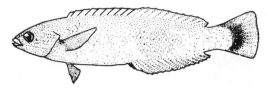

Señorita
(Illustration courtesy of California Department of Fish and Game)

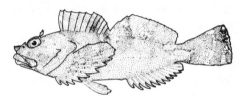

Cabezon
(Illustration courtesy of California Department of Fish and Game)

The fronds and stipes of kelp are quickly colonized by opportunistic larvae of animals set adrift in the plankton. A blade can be completely covered by attached forms of life within three weeks. Kelp does not get weighed down by heavy encrusters like mussels and acorn barnacles because it grows fast and sheds the ends of its blades long before those creatures can grow to appreciable size.

Lacelike mats of bryozoans and multitudes of hydrozoans gain a fast foothold on the fronds. In turn, these newcomers fall prey to the sea slugs, the *nudibranchs*. Basically unshelled snails, the nudibranchs are among the most colorfully flamboyant members of the mollusks. Their bright colors and conspicuous markings would seem to make them easy targets for potential predators but, like the butterfly, their gaudiness also advertises an acrid, toxic, and unpalatable mouthful. Some nudibranchs who feed on stinging jellyfish can assimilate the sting cells intact and transfer them to defensive positions on their own bodies. The only creatures they have to fear are near-relatives.

The **frost spot,** *Corambe pacifica,* is the most voracious predator of the bryozoan *Membranipora.* No more than ten millimeters long and nearly transparent, it rasps into the bryozoan colony and sucks out the soft parts. The narrow, coiled, white bands of its eggs are easy to spot on the kelp. A related species, *Doridella steinbergae,* also feeds on *Membranipora,* often in vast numbers.

The sea slugs range in size from minute forms that are able to live between grains of sand to the **California black sea hare,** *Aplysia vaccaria,* which can reach a weight of 15.9 kilos (35 pounds) and a length of seventy-six centimeters (thirty inches). It feeds almost exclusively on **feather boa kelp,** *Egregia,* which grows close to shore off Central and Southern California.

Smaller, and much more common, the **brown sea hare,** *Aplysia californica,* browses in the kelp canopy, among other places. It has been reared successfully in the laboratory. Because its nerve-cell bodies are so large, so distinctly marked, and so few, they are used extensively in nerve and behavior research.

The holdfast of the kelp, with its interwoven network of tendrils and crannies, is a microenvironment in itself. The tendrils create hidden voids and provide enormous surface area. A large holdfast can harbor more than one hundred thousand organisms that can be seen with the naked eye. Add in microscopic critters, and the count climbs into the millions. Crustaceans, mollusks, polychaete worms, anemone, brittle stars, sponges, and tunicates are all represented. Most attach to surfaces, but a few burrow in: the **gribble,** *Limnoria,* for example.

Holdfasts are nurseries for the juveniles of many species. A single giant kelp holdfast may contain up to fifty miniature abalone of one to three millimeters in length. Eight species of abalone, *Haliotis,* occur off California and are found from the intertidal zone to 150 meters deep. It takes eight years for an abalone to reach maturity.

In recent years, climatic changes in the southern Pacific Ocean—unusual extensions of a seasonal warming effect called "El Nino"—have caused heavy seas and brought warmer water to the California coast, with ruinous effects on the kelp beds. But the decline of the kelp beds started long before these warm-water intrusions. In the 1950s, the demise of the kelp beds was attributed to the explosive increase of two species of sea urchin, *Strongylocentrotus franciscanus* and *S. purpuratus.* Sea urchins attack kelp holdfasts. Once the plant tears loose from the bottom, it cannot reattach. Why had the sea urchins increased in numbers and what sustained them in vast numbers after the kelp was gone? Some suggest the urchins are capable of living on dissolved organic matter from sewage. Others think the diatom and microalgal films that grow in place of kelp on the bare rock supply enough sustenance to see the urchins through.

The urchins are the natural prey of the **sea otter,** *Enhydra lutris.* The sea otter was hunted to near-extinction at the turn of the century. Now protected, its reappearance has, in the areas lucky enough to be visited by otter herds, coincided with precipitous declines in urchin populations and a resurgence of kelp. As urchins are brought under control, kelp prospers. That's fine if you harvest kelp. If you hunt abalone, the return of the otter is bad news because the otter finds it a tasty change from a monotonous urchin diet. Otter take one abalone for every twelve urchins.

Life Among the Rocks

Under water, the Pacific Coast is far from a flat expanse of rock. It has steeply sloping cliffs, narrow canyons, deep ledges, and rock overhangs that all provide a base for an extraordinary amount of attached life. Walls and overhangs are carpeted with brightly colored encrusting and tube-shaped sponges, bryozoans, tunicates, colonies of mussels, and red or brown algae either in mixed patches or in masses of a single dominant species. **Northern palm kelp,** *Pterygophora,* and **blade kelp,** *Laminaria,* dominate large tracts in the colder north-coast waters.

As you might expect, the ledges, cliffs, and walls attract a wide variety of fish. In Southern California, where the attached life includes the soft coral sea fans and sea whips of warmer water, the gaudy **garibaldi,** *Hypsypops rubicundas,* as bright orange as a goldfish, is every bit as attractive to the eye as is the queen angelfish of the tropics.

Reams have been written about Pacific-coast gamefish, yet, as on the East Coast, they offer little opportunity to gain insight into their lives by direct observation. It is the smaller fish, generally homebodies who are territorial and endowed with elaborate behavior patterns, that can be observed profitably and their ways eventually understood.

Consider the **blacksmith,** *Chromis punctipinnus.* When courting time comes, the male builds a nest in a cave-shaped crevice between cobble and boulder in the ledges. He cleans the cave roof, then pushes the egg-laden female into the site. She attaches the eggs to the roof with an adhesive filament; just how isn't known. Then she leaves, abandoning eggs and companion. The male backs into the cave and fans the eggs with his tail. He ferociously drives off all intruders for, should he leave the nest, the eggs would quickly be devoured by other blacksmiths, senoritas, garibaldis, or sheepshead patrolling the area.

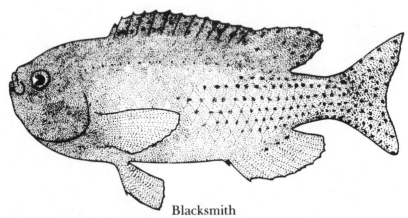

Blacksmith
(Illustration courtesy of California Department of Fish and Game)

While on guard duty, the male changes from his normal steel gray to a mottled white with two distinct dark bands near the eyes. Nest-guarding may not be induced the same way in all fishes that do it, but

A garabaldi guards its nest. The surrounding rock has been bared by sea
urchin. *(Photo by Ron Church)*

in the much-studied **three-spined stickleback,** *Gasterosteus aculeatus,* the ventilating fin motions are triggered by a chemical stimulus emanating from the eggs.

The **blackeye goby,** *Coryphopterus nicholsi,* behaves much the same as the blacksmith after egg-laying, but before, he performs a courtship display. He rises straight up off the bottom, then settles back, spreading his dorsal and pelvic fins before the female.

Courtship displays are probably much more common among fishes than we suspect. The hormone changes in breeding season make the males aggressive, modify skin color, and set the trigger for sexual action. Whether the sexes first build a nest or simply shed eggs and sperm into the water, timing is crucial. Something must set things off. It can be body language—as simple as increased swimming speed or as complicated as the elaborate rituals of the cichlids, so highly favored by aquarists for their flashy ways.

Little fish often develop odd associations. Deep in Monterey Canyon, the large **California king crab,** *Paralithodes californiens,* inadvertently provides shelter for the eggs of the **blacktailed snailfish,** *Careproctus melanurus.* Members of the Moss Landing Marine Lab were astonished when hatchlings began to emerge from the gill cavity of a crab that had been caught in a bottom trawl. They counted over a hundred eggs in the gills, almost all of which hatched within twelve hours.

The same thing has been seen in the **Alaskan king crab,** *P. camtschatica,* and in *Lopholithodes forminatus,* both deep-water inhabitants. The gill cavity of one *Lopholithodes* was stuffed with over four hundred fish eggs, which probably impaired the crab's respiration. The female snailfish deposits the eggs with a long oviposter that she inserts into the crab's gill cavity. Biologists are hard-put to say whether this is an example of *commensalism,* a close association between two species where only one derives any benefit, or short-lived parasitism, where the crab is only temporarily harmed.

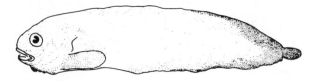

Blacktail snailfish
(Illustration courtesy of California Department of Fish and Game)

THE ATLANTIC COAST

Rockbound from Avalon Peninsula in Nova Scotia to the shores of Rhode Island, above Cape Cod the sea floor is a patchwork of bare rock, ledges, loose boulder and cobble, and sand and silt—the remnants and handiwork of glaciation.

The water and its life are chilled by the southerly flow of the Labrador Current. Water temperatures rarely exceed 10° C except in coves and inlets warmed by summer sun, and fall to just above freezing in winter.

Seaweeds dominate the coastline and subtidal waters. The shoreline is rocky, indented, and enormously long, the result of a rising ocean and scoured, subsiding headlands. The tidal range is large: 3.3 meters from Boston Harbor to Penobscot Bay in Maine, then increasing to 6.4 meters at Eastport, Maine.

A note of caution to the underwater naturalist here: These sizable tidal differences create swift currents over shoals between the many islands and outcrops in these waters. Plan your explorations to coincide with slack water—avoid full flood and ebb. Never leave a boat unattended (Maine folklore is filled with stories of men drowned when a rising tide and swift water carried off their skiffs and left them stranded).

Slack water can end and a current can begin before you realize you cannot swim against it. If you are close to the boat, using scuba gear,

The sand on Sand Beach, Maine, in Acadia National Park is mainly shell fragments with little or no quartz.

and have a reasonable air supply left, go to the bottom and pull your-self up-current hand over hand by grabbing on to seaweed. Better yet, set a float out on a long line from the boat before going in the water, and don't stray beyond the end of the tether. If neither works, simply float and wait for the boat to up-anchor and fetch you.

Surface current often runs stronger and sooner than bottom cur-rent. When you see all the seaweed begin to bend in the same direc-tion or all the fish lining up one way, get out of the water. Currents above four knots are not uncommon, and you cannot swim against that even with the largest flippers made.

The brown algae of the East Coast shorelines are called *wracks* and grow in distinct zones. **Channeled wrack,** *Pelvetia canaliculata,* oc-cupies the highest fully submerged zones. **Knotted wrack,** *Ascophyllum nodosum,* and **bladder wrack,** *Fucus vesiculosis,* are mid-tidal species and seldom found together. Knotted wrack often grows on stones and can eventually become so buoyant that it floats them. Some cob-ble beaches have been built up from waves tossing rock and wrack ashore, leaving the wrack to rot and the stones to stay. **Serrated wrack,** *Fucus serratus,* has no air bladders and grows below the low-tide line.

Below tide-exposed bottom, the zones of seaweeds depend on surge, currents, and light. **Winged kelp,** *Alaria esculenta,* grow from low tide to six meters below in rough water or swift current. *Laminaria* species, **oarweed,** with big, broad, single blades that can be from two to seven meters long, cover the bottom in dense leathery masses from the low-tide line to deep water.

The holdfasts of oarweed grow close together. Very little open space exists between them, but what space there is provides hiding places for large invertebrates and bottom-dwelling fish.

The keystone alga grazer is the **green sea urchin,** *Strongylocentrotus droebachiensis.* Unchecked, green sea urchins can decimate seaweed beds, and in many areas great numbers cover the rock bottom. Their major predator until recently was the **American lobster,** *Homarus americanus.* In the last few years, an export market for urchin roe has developed with Japan, and they are being harvested in enormous quantities.

Either the green sea urchin or its close relative, the **purple sea urchin,** *Arbacia punctulata,* make fine additions to the saltwater aquar-ium. You can watch the actions of the tube feet, the spines, the clawlike pinchers on their upper surface, and the workings of the five

Holdfasts of oarweed, Maine coast

Green sea urchin on a rock coated with coralline algae. A cluster of anemones cling to a rock behind it.

teeth that line the mouth and scrape algae from the bottom. You can extract eggs and sperm from ripe adults and, with a low-power microscope, watch the fertilized eggs divide and grow. Their development follows a precise and classical embryological pattern.

During late summer, gather a few large specimens of *Arbacia*. The conservative way of extracting the eggs and sperm is by mild electrical shock. Apply a ten-volt AC current (from a household bell transformer) to any two points on the urchin, using lead electrodes. The urchin will immediately begin to shed eggs or sperm if ripe. The urchin is not harmed, and you can repeat the procedure either immediately or at a later time.

Once you have separate quantities of both egg and sperm, dilute the eggs in a small cup of seawater and add a few drops of freshly diluted (about 1:100 seawater) sperm. As the eggs fertilize, they will sink. At 23° C, first cleavage will begin in 50 minutes, second cleavage in 78 minutes, and third cleavage in 103 minutes. Division continues, first forming a single-layered hollow sphere of cells—the *blastula*—then, by inward folding, a two-layered *gastrula*, which develops further into a free-swimming *pluteus* larva. The blastula forms in seven to eight hours, the *gastrula* in twelve to fifteen hours, and the *pluteus* larva in twenty-four hours. You can feed the pluteus larva a diet of diatoms (*Nitzschia*), and within two weeks it will metamorphose. The rate of development depends on temperature, but don't push it: 30° C and above is lethal to the larva.

Other than oarweed, the most striking aspect of the New England underwater rock is the hard incrustations of pink and white algae. These calcified plants are so extensive that you can get the impression the rocks are naturally pink.

Found from New Jersey to the Arctic, species of *Lithothamnium* form a bumpy pink-to-red crust over virtually any hard surface they happen to find. Thinner, more like a coat of paint than a coat of rough plaster, pink *Clathromorphum* widely encrust rock north of Massachusetts. Small shrubby growths of *Corallina officinalis* also cover the rocky bottoms. Its segmented branches are coated with a stony coating of calcium-magnesium carbonate. Alive, the plant is reddish purple; dead, it is chalk white.

Sponges are common: the yellow-green **bread crumb sponge,** *Halichondria panicea;* white, warty *Melonanchora elliptica;* pale *Pellina sitiens.* All have wide openings through which the water taken in by minute pores in the walls is expelled.

Rock and mussel encrusted with the algae *Lithothamnium*

You are sure to come across the pink-white **finger sponge,** *Haliclona oculata,* which grows in clumps like a small bush and can stand as much as a half-meter high.

A bright yellow species, *Cliona celata,* grows on mollusk shells, and slowly eats into them by dissolving the calcium carbonate in the shell. Clam shells riddled with hundreds of holes are often found on the beach, the work of these and other boring sponges.

Although sponges are whole animals whose cells form differentiated parts, they also behave like colonial associations and have a curious regeneration capability. If you take a piece of sponge and grate it lightly through cheesecloth into seawater, then pour some of the suspended cells into a shallow dish, the cells will not only settle to the bottom, but will make their way to one another and reorganize. Mix the cells from two species and they will sort themselves out. What happens if you take two individuals of the same species but collected from widely separated areas?

For sheer obscurity among the general public and even among first-year biology students, the tunicates rank high. One is hard-put to understand what they are doing in the same phylum as the chordates, the kin of fishes and mammals. You cannot tell until you look at their free-swimming tadpole larva. It is equipped with a *notochord* and a *tubular nerve cord:* forerunners of the vertebrate spinal cord and backbone. The larvae settle out, attach, then lose their advanced status by metamorphosing into simpler sessile forms. Some are colonial and encrust the rock: the **golden star tunicate,** *Botryllus schlosseri,* for example. Others are solitary: the **sea peach,** *Halocynthia pyriformis,* is a single peach-sized creature that extends two large siphons, one to draw water in, the other to expel it after filtering out food and taking up oxygen.

There are so few places to burrow that casting a hard casing makes a sensible alternative. The tube worm *Filograna implexa* forms colonies of twisted hard tubes six centimeters in length that look like open stiff pile carpet. On seaweed, you will often find the small counter-clockwise-coiled tubes of *Spirorbis borealis,* a serpulid worm, and netlike patches of the **sea lace,** *Membranipora pilosa,* a bryozoan.

If you snip off a piece of sea lace and look at it under a low-power microscope, you will find each cell structure is an individual animal whose mouth and surrounding tentacles will be busily sweeping the water for a plankton meal. Another bryozoan, *Bugula turrita,* forms dense tufts of white-to-orange colonies that collectively look like the bushy limb of a miniature pine tree. These, too, are worth a closer look.

Stony surfaces make good browsing grounds for mollusks. Small chitons and limpets cling to the surfaces and feed on the layer of diatoms and microalgae that coat the rocks. They don't appear to move, but each one grazes down its surround, preventing anything substantial from gaining a foothold.

Whelks, who are active carnivores and scavengers, range from thumbnail size to the length of a hand. Their empty shells are quickly recycled by hermit crabs. *Pagurus acadianus.* One of the larger hermits on the New England coast, it will, at first, retreat into its home when picked up, but, if you hold on too long, will deliver a pinch you will not soon forget.

The commonest of all the true crabs along these coastlines is the **rock crab,** *Cancer irroratus.* An aggressive predator, it feeds on smaller crabs, urchins, snails, and clams—about anything it can capture. If you put one in your aquarium, you will find it is a digger as well and will plow up the bottom from one end to the other.

Adult male rock crabs can grow a shell thirteen centimeters wide. Both claws are about equal in size and shape. If you look closely while they are feeding you will see that a pair of smaller appendages closer to the mouth do the actual feeding. The big ones do the tearing, cutting, and splitting.

In the spring, you may find crabs with large egg masses underneath the *apron,* a flap of shell on their undersides. They breed from late summer through fall and the female carries the eggs until they hatch in late spring of the following year. The hatchlings start as a larval form called a *zoa* and molt often. With each molt they change progressively through several intermediate stages before becoming a small crab. They take about two years to reach breeding age.

The **Jonah crab**, *C. borealis*, grows to be larger than the rock crab, but occupies a narrower geographic range, being bound to year-round cold water. Although both the rock crab and the Jonah are loners, divers have come across mass aggregations of both species. Grouped together on rocks, freshly molted adult crabs have been seen in numbers exceeding two hundred individuals per square meter in 6° C water in December. Follow-up surveys in mid-January in water just above freezing, found they had migrated to nearby sand, burrowed in, and were inactive.

Neither the rock crab nor the **American lobster,** *Homarus americanus,* are exclusively rock-bottom dwellers. Even though the "Maine" lobster is mentally entrenched as a Down East symbol, its geographic range extends from Canada to the mid-Atlantic states. *Homarus* is the object of intense fishing wherever it is found. Migrating offshore into deep water in winter and returning inshore in spring and early summer, fishery biologists estimate that ninety-five percent of all inshore seasonal migrants of legal size are captured. Maine has recently decided to increase the legal-catch carapace length from 3¼ inches to 3⁵⁄₁₆ inches by 1992, thus increasing the average catch weight from 1⅛ pounds to 1¼ pounds, and giving it one more season to breed.

Each northeastern state regulates the catch and who can do the catching. In Maine, local custom goes well beyond the law, dictating who can get into the business and at what pace. The competition for trapping space in Maine has led to a supralegal allotment system. Each lobstering community has its own traditional grounds and does not welcome pots from fishermen in neighboring communities. If you set traps in someone else's water, the first warning is a knot in the float line. Second warning is a cut line and a lost string of pots.

Scuba diving for lobster is legal in some states and not others. It is forbidden in Maine and open to residents in Massachusetts. No state allows taking a female in berry (with eggs). Divers see and take much larger lobster than do commercial fishermen using pots. Above 4.5 kilograms (10 pounds), a lobster is invariably male and, for the sake of the lobster population as a whole, taking one this size helps, for it will prey heavily on smaller lobster. Big lobster, between 4.5 and 9 kilos, were once relatively common in waters eighteen to thirty-six meters deep off the Northeast Coast, but the numbers are dwindling.

Given the prices fetched by lobster, their breeding has been intensely studied over the years, but not crowned with practical success. Lobster molt frequently when young but, unfortunately, are cannibalistic and cannot be raised in aggregates. Growing them individually is an expensive business because they take six years to reach marketable size. Juveniles have been raised by state hatcheries through initial molts and released, but there is no way of telling whether they ever reach maturity. One attempt to breed a rare blue form ran out of funds before a large-enough population was established to conduct conclusive experiments.

The **purple star,** *Asterias vulgaris,* is among the key bottom predators. Not many attached mollusks can withstand its attack. The **sun star,** *Crossaster papposus,* unlike the normal five-armed stars, has eight to fourteen arms, and like the purple star, can grow to an overall diameter of thirty centimeters. It feeds on other starfish.

The curious **basket star,** *Gorgonocephalus arcticus,* either balled up or fully expanded, is an oddity among echinoderms in particular and animals in general. Its tangled and coiled white tentacles radiate from five arms like branches from a flexible tree. During the day it is found clinging to underwater growth. Usually more than one are together in a tangled mass. At night it expands its arms out into an undersea net, fishing for plankton.

Its relatives, the sea cucumbers of the North, do much the same thing. Soft-bodied, they cling to rock and extend the ring of tentacles around their mouths. *Cucumaria frondosa* has ten tentacles that extend out over thirty centimeters, capturing plankters among its branchlets, which it then sweeps into its mouth.

Fish of the Atlantic Coast

New England waters are among the most productive fisheries in the world. The commercially important fish families include herring, hake, cod, flounder, sole, mackerel, tilefish, weakfish, and rockfish. Either in schools or aggregates near the cobbled bottoms, they are harvested by seining, trawling, or longline.

Among the cracks and crevices, along ledges and between boulders, the more cryptic and solitary species reside. Sculpins and ravens, with their broad spiny pectoral and dorsal fins, are camouflaged in blotchy, variegated gray and brown. They blend into the background and lie in wait for a passing meal. But the vividly yellow **sea raven,** *Hemitripterus americanus,* provides another exception to nature's rules. Neither fast nor known for maneuverability, its stomach contents often include shrimp, mussels, worms, ascidians, crabs, amphipods, and more, suggesting they go foraging as well as lying in wait.

Hiding in weed or clinging onto a rock with its sucker disk, the **lumpfish,** *Cyclopterus lumpus,* munches on a variety of invertebrates. It breeds in shallow water in spring and early summer. The male guards the eggs after the female abandons them. Seals hunt them, but to what extent isn't known.

You will find smaller fish among seaweed fronds and in crevices. In some places it seems as though there is a **rock eel,** *Pholis gunnellus,* under every flat stone. **Sea snails,** *Neoliparus atlanticus* and *Liparis liparis,* are small and have a sucker disk much like the lumpfish's. Look for them on rock and weed stipes. They make a fine addition to an aquarium.

Ambush is the stock-in-trade of the **American goosefish,** *Lophius americanus,* better known to the fisherman as the *allmouth,* an apt appellation to any who have seen that apparatus in action. It dangles a fleshy blob on a forward dorsal spine in front of its enormous head, and any hapless creature attracted to it is suddenly sucked into the allmouth's maw by the swift intake of water as it opens a semicircular set of jaws capable of engulfing a basketball.

Goosefish are found along our northern coast as far south as Cape Lookout, North Carolina, from twenty to six hundred meters deep. They spawn in spring, summer, and fall. Divers off New Jersey have seen congregations of them in late winter and early spring when the water temperature is 4° C. Whether these gatherings are a prelude to mating isn't known.

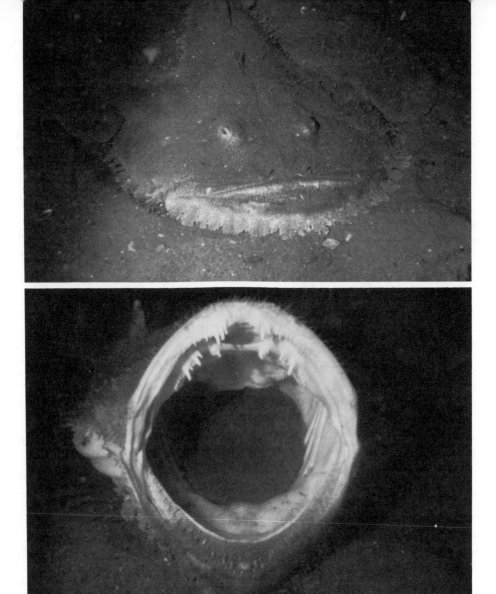

This is why *Lophius americanus* is also called the "allmouth."

The prize for fierce appearance among fishes goes to the **wolffish,** *Anarhichas lupus.* Its overall shape reminds one of a blenny, albeit a mighty big blenny, for it can run to two meters in length and weigh eighteen kilos (forty pounds). When it opens its mouth, the blenny analogy is immediately abandoned for the image of an angry Doberman. The teeth, though ferocious, are used exclusively to grind urchins, crabs, and mollusks. This does not mean the wolffish can be handled with impunity. Caught on hook and line, it snaps viciously and effectively at anything within reach.

Fish Diversity

Of the twenty-two hundred species of American finfish, over fifteen hundred are marine. Of those, only eighty-one are found in both Atlantic and Pacific, and those are mainly the large pelagic wanderers.

Of the 163 families of marine fish in U.S. waters, just over seventy families have three or less species representing them. Those families with many species in these waters may owe their diversity to their special surroundings. Whether due to the beds of kelp or the craggy, pocketed bottoms, the Pacific Coast harbors 138 of the 196 species of sculpins and scorpionfishes who hunt from ambush and prefer ample cover.

The American lobster, *Homarus americanus,* hides by day and hunts by night. In large lobsters, a third of the body weight is in the claws. *(Photo by Jerome Prezioso)*

Some family differences have no rationale other than wide geographic separation. The preponderance of lefteye flounders are found on the Atlantic Seaboard, righteye flounders on the Pacific.

Diversity among fishes increases in warmer waters. Probably because our Atlantic and Gulf coasts are close by the Caribbean Sea, the Atlantic has the bulk of the typical tropical families; cardinalfishes, jacks and pompanos, snappers, mojarras, grunts, drums, butterflyfishes, damselfish, wrasses, parrotfish, mullet, barracudas, jawfishes, clinids and blennies, gobies, butterfish, triggerfishes and filefishes, puffers, and porcupinefishes. Although tropicals are not well represented in our eastern Pacific waters because of a lack of habitats, the diversity of the southwestern Pacific tropics is far larger than that of our tropical Atlantic.

7

Coral Reefs and Tropic Seas

Conjure up a vision of a bright blue sky above a warm turquoise sea. Edged by the glaring white sand of a palm-lined beach, green water extends offshore, then abruptly changes to deep blue beyond the breakers. This is the realm of the coral reef. The life within it is the richest and most beautiful in the sea. No more diverse and prolific region can be found underwater. Once you have visited, you will be drawn back time and again, for few places can rival its attractions.

Although the southern tip of the continental United States lies above the Tropic of Cancer, the Florida Current bathes the eastern side of southern Florida in waters warm enough for reefs to flourish. From Fowley Rocks in Biscayne Bay to the Dry Tortugas, corals grow fast enough to sustain and enlarge the limestone monuments that harbor so many creatures.

The Keys, its reefs, Florida, and the Bahamas all sit atop a broad limestone base shaped by reef growth eons ago, the product of emergence, erosion, and resubmergence over geological time. The Florida Reef Tract, as the underwater region is known, lies in the ocean side of the Florida Keys, eight to eleven kilometers offshore. To the north, off Key Largo, lies John Pennecamp Coral Reef Preserve, which extends from Turtle Reef at its northernmost end to Molasses Reef twenty-nine kilometers to the south. Between the two lie well-known diving locations that draw hundreds of thousands of scuba divers and snorklers each year: the Elbow, Grecian Rocks, and French Reef. Grecian Rocks is shallow and suitable for snorkling.

South of Pennecamp Park, Alligator Reef, Sombrero Key, Looe Key, and American Shoals form a sea chain that parallels the Keys as far south as Key West. Midkey reefs are not as well developed as those either to the north or south; cold winter water from Florida Bay discourages their growth. From Marathon Key southward, off-shore reefs are much like those in the Caribbean, with well developed fore-reefs, and buttress and breaker zones.

STRUCTURE OF THE REEF

Reef-building corals thrive in shallow tropic seas where the water is clear, salty, sunlit, and warm. Although some species of coral grow in temperate waters, they only grow fast enough to create reefs in waters where the temperature does not fall below 18° C year-round. Water temperatures around 24° C are ideal. Growth can occur in water as low as 20° C and as high as 36° C, but much less vigorously. Thus, the majority of the world's reefs lie between the Tropic of Cancer and the Tropic of Capricorn.

Corals grow best in waters whose salinity is thirty-four parts per thousand—open ocean. The waters must be free of silt, although a few corals can tolerate some silt and low salinity well enough to grow close to shore.

The coral reef is a living veneer of animals actively growing on a base of the remains of earlier colonies. Over geological time, large land masses have been created by these endless replications. Islands and continental margins have been shaped by their ceaseless activity over millions of years.

Charles Darwin was the first to describe classic reef forms: fringing reefs, barrier reefs, and atolls. *Fringing reefs* build on and extend out from a rocky shoreline. *Barrier reefs* build up parallel to the coast in a shallow sea, often at a considerable distance from land. For example, the Great Barrier Reef off northeastern Australia parallels its coast-line for nearly 1600 kilometers and lies 50 to 240 kilometers offshore. *Atolls* are a ring of reefs in open ocean surrounding a shallow lagoon. Darwin speculated that atolls begin as fringing reefs around a subsiding volcanic cone. As the mountain of basalt slowly sinks over thousands of years, the corals grow upward fast enough to keep pace with the submergence. The net result is a circular chain of reefs. Similarly, he reasoned, barrier reefs form by slowly subsiding coastland.

Most of Darwin's ideas were drawn from Indo-Pacific examples and without the benefit of later findings of geologists. He classified Floridian and Caribbean reefs as fringing reefs, but their origins, like those of many Pacific reefs, have proved to be more complex than he thought. He did prove correct about the origin of atolls.

West Indian reefs have a typical architecture, but it is often highly modified from one place to another by underwater topography, current flow, wave action, suspended sediments, and a host of other extenuating circumstances. The best example of an archetypical West Indies reef is found along the north shore of Jamaica. A breaker zone runs parallel to the coast where the sea crests and spills over into the shoreward shoals beyond the reef flat; it is so shallow that it may be partly exposed at low tide. Further landward lies a rubble field: coral debris thrown shoreward by waves. Very little coral grows there, but patches of algae and expanses of the green colonial anemone *Zooanthus* prosper; so much so that the region is called the *Zooanthus* zone. Between land and the *Zooanthus* zone lies the lagoon, of which more will be said in the following chapter.

The crest of the reef at the breaker line is dominated by **elkhorn coral,** *Acropora palmata,* to such an extent it is called the *palmata* zone. These corals take the full brunt of the surf, a place where few other corals can grow.

Seaward, below the *palmata* zone, lies a gently sloping area filled with loose rubble and occasional stands of elkhorn coral. Farther below, coral-covered mounds, called *spurs* or *buttresses,* rise from the bottom and extend seaward perpendicular to the reef crest. These mounds can be thirty meters long, three to ten meters high, and are rarely narrower than three meters. Between their parallel walls lie narrow coral sand canyons one to two meters wide, called *grooves* or *channels.* They are often close enough that corals from adjacent buttresses extend out and roof over portions, turning them into tunnels or caves. How this *spur-and-groove* (or *buttress-and-channel* as it is alternatively called) arrangement comes about isn't known, but the likely explanation lies in the great volume of sediment formed in the *palmata* zone. The return water from the breaker zone carries the coral sand seaward along the bottom through these channels.

Heaps of staghorn coral in a mixed stand of corals

Lettuce leaf coral, *Agaricia agaricites*

If the channels were not there, the sediment would soon scour and choke the buttress reefs out of existence.

Although the buttresses are constructed by the growth, death, and compaction of a great many coral species, the massive wall builder *Montastrea annularis* does most of the work. Below the buttress-and-channel zone, which typically ends at a depth of eight meters, a tract of sand and coral rock begins and gently slopes seaward for thirty to ninety meters to a depth of fifteen meters or so. Immense heaps of **antler** or **staghorn coral,** *Acropora cervicornis,* litter this bottom.

Around a depth of twenty meters, the bottom sharply rises, staghorn coral disappears, and *M. annularis* codominates with *Agaricia agaricites,* **lettuce coral,** whose leafy plates cover wide expanses of this fore-reef. The seaward side of the fore-reef is the edge of the escarpment, the "wall" or "drop-off" in diver's parlance. As the gradient steepens, hard corals give way to soft corals—sea whips and sea fans.

Other places show other arrangements. Where the seaward slope is long and waves are moderate, *Acropora palmata* may dominate the breaker zone both on the seaward side and well inshore of the surf line. Seaward of the elkhorn coral, heaps of staghorn coral often intermix with it and the buttress zone is confined to deeper water. In other areas, clear-cut zonation is muted or nonexistent.

Corals and Associates

West Indies and Florida reefs do not always follow predictable sequences. Lagoons may either be barren or filled with small reefs surrounded by limey sands, grass flats, or hard bottom. In deeper water, fifteen meters or so, the bottom may not follow the Jamaican example but simply be littered with mixed patch reefs separated by narrow sediment beds. These mixed patch reefs go through cycles of growth and decay that leave them pockmarked with holes, interconnecting passages, overhangs, and mini-caverns.

THE CORAL ANIMAL

Corals are kin to the sea anemones. Hard coral polyps are ringed with six or a multiple of six (usually twelve or twenty-four) tentacles that sit atop a contractable sac. The tentacles are lobed with stinging cells. When chemically stimulated, they discharge a barb filled with a paralyzing toxin that stuns and holds microscopic prey.

The tissue in the sac is folded and follows the outline of the skeletal cup, which the polyp builds by secreting calcium carbonate at the floor and sides of the cup. Within the sac, the tissues specialize in digestion and reproduction. Threadlike filaments with powerful digestive capabilities also originate here and extend into the interior of the sac and out through pores in the walls into the spaces between the polyps.

Certain hard corals have a symbiotic relationship with chlorophyll-bearing algae, called *zooxanthellae*, that live within the corals' tissues. They convert sunlight into nutrients, thus helping the coral to prosper. In the Caribbean, most reef-building corals are limited to depths no greater than twenty-seven meters, although slow-growers exist down to sixty meters.

The zooxanthellae are stored inside the cells of the coral polyp; their numbers are in proportion to the amount of light the coral receives. If the coral is blocked off from light for a few months, it will expel the zooxanthellae, presumably to lighten the metabolic load these tiny cells impose in continuous darkness.

Zooxanthellae are passed on to future coral offspring through the *planula,* the free-swimming larval stage of the coral. Specific species of zooxanthellae have adapted to specific corals.

Considerable controversy rages over the role of zooxanthellae as a source of food for corals. They are not digested directly; that is, the coral does not farm and harvest them, but they do supply soluble nutrients to the coral and use waste products of coral metabolism such as ammonia and carbon dioxide. Zooxanthellae also aid reef-building. They consume phosphorus, which otherwise would inhibit calcification. Calcium carbonate is secreted as the mineral *aragonite* by the polyps onto an organic interlayer that provides sites for mineral growth. Organic phosphorus compounds stifle mineral formation.

A coral colony does not continue to grow indefinitely, but reaches a maximum size. The hemispherical brain corals rarely exceed two meters in diameter. Elkhorn coral rarely fans out wider than three meters. Mounds of *Montastrea* become more and more bumpy as they expand, and finally evolve into a series of separate lumps. *Montastrea* also change form with depth, becoming more platelike as the light fails.

A Nassau grouper hovers beneath the outsretched branch of elkhorn coral.

The growth rate of corals varies from ten to twenty-five centimeters per year in length for staghorn coral, to one centimeter in diameter per year for brain and finger corals. The remains of shipwrecks on reefs have helped establish average growth rates; corals attached to the ships' steel obviously can be no older than the date of the vessel's demise.

Corals are excellent microcarnivores, catching just about everything that settles on them. They trap plankton by stinging them or entangling them in mucus. Polyps feed at night when plankton rise to the surface from the deeps. Corals with large polyps trap more plankton and have fewer zooxanthellae than do smaller-polyped corals.

The first impression of a coral reef is one of rampant animal life. Where are the plants to feed them all? Very little green shows anywhere, and tropic waters don't hold enough plankton to provide for such abundance. To measure the net productivity of the reef—the excess of plant growth over animal uptake—biologists measure the increase in oxygen over the reef during the day and compare it to its decline during the night. The day-to-night change can be converted into *net photosynthesis,* which is expressed as the amount of carbon "locked up" in converting carbon dioxide into life. The figures are staggering. Open-sea productivity in the tropics averages forty grams of carbon per square meter per year, but the waters around the reef may be as high as thirty-five hundred grams per year. Where is all the plant life located?

Zooxanthellae account for some of it. Encrusting algae growing on dead coral and among the grains of the coarse coral sediments account for most of the rest. Constant close cropping by fishes and urchins keep the algae inconspicuous. If you remove the major reef-grazer, the **spiny sea urchin,** *Diadema antillarum,* algal mats will quickly appear. Unchecked, these mats will overgrow sessile creatures, including the corals, and destroy them.

Life Among the Coral

Corals suffer continual predation. The parrotfishes and the surgeonfishes nip off live coral in their search for algae. In parrotfish, the teeth have fused into a beak that can scrape a sizable chunk from a coral head. You will see the white scars on almost all coral surfaces. The ingested coral is crushed, the organic material absorbed, and the remains defecated as a fine silt. This silt settles and eventually consolidates into a hard bottom sediment. The coral's scars either heal over with new coral growth or become the points of entry for colonists such as boring sponges, tube-building worms, or algae.

Dead coral is continually scraped by urchin, *Diadema* for one, and *Echinometra viridis* for another. Scraping prevents new organisms from gaining a foothold, but also wears depressions in the coral that weaken their structural integrity. Boring sponges dissolve both living and dead coral: *Siphonodictyon coralliphagum,* a yellow, tubed type, and *Cliona delitrix,* a red incruster often dotted with *Parazoanthus* anemone, burrow, riddle, overgrow, and kill large corals. They also weaken the base of the coral and destroy it by detaching the main body from the base.

Most tropic sponges are benign, simply using the dead coral as a foothold. In deep water, *Mycale laeris* can be downright helpful. Growing in a flattened form under the **leaf coral,** *Agaricia agaricites,* it crowds out boring forms and cushions the coral in an upright horizontal position if its small stalk cracks off from storm surges.

A vase-shaped sponge,
probably *Niphates* sp.

Sponges add bright color and contrast to the reef scene and harbor their own commensals and parasites. The interstices of the common **loggerhead sponge,** *Speciospongia vesparium,* a barrel-sized black mass with a flattened top and large circular openings, provides a home for a great many porcellanid crabs and pistol shrimp. The **snapping shrimp,** *Synalpheus,* is one such tenant. Like other pistol shrimps, it has one enlarged claw that slams shut with an audible click.

The **mantis shrimp,** *Gonodactylus,* sometimes lives in the loggerhead sponge, but is more commonly found in coral crevices where it lies in wait for its prey. Armed with two unusual forelimbs, it is a formidable predator. The forelimbs are hinged and fold back like those of a praying mantis. The inside edges of the forelimbs are lined with sharp spines that initially impale, then snap shut like an iron maiden, sometimes cutting the victim in two.

The fronts of the forelimbs of these species can be shaped either like a spear or a hammer. Their strike is one of the fastest blows in nature, taking only four to eight milliseconds to deliver. In those shrimp with limbs built for smashing, the forelimbs can split a crab with a single thwack. *Hemisquilla ensigera,* twenty-five centimeters in length, can deliver a punch with a force nearly equal to a small-caliber bullet. They are nicknamed "thumb busters" for good reason.

Totally passive and apparently defenseless, what protects sponges from depredation? Neither their fibrous inner structure nor the spicules that frame their inner walls are decisive deterrents. The answer lies within the fleshy tissue, which ranges from terribly distasteful to downright poisonous. Both the massive brick-red **touch-me-not sponge,** *Neofibularia nolitangere,* and the bright-orange-to-red *Tedania ignis* are so toxic that a diver will develop blisters from handling them. Most tropic sponges are not dangerous to the touch, but enough red ones are that it's wise to avoid them until you know your species well enough to tell the innocuous from the harmful.

The surface settlers on the massive corals include the **Christmas tree worm,** *Spirobranchus giganteus,* one of the many fan and feather-duster worms whose bright gills sweep the waters for plankton. When disturbed, it retreats into its tube in a wink. As the coral grows, these polychaetes add new tubing, giving the impression the worm burrows in, which it does not. **Date mussels,** *Lithophaga,* bore into coral, living or dead, until all that is seen of them is a dumbbell-shaped opening. If the coral is alive, they must continually bore outward to keep from being overgrown.

. **Gall crabs,** settling on coral, dig their own specialized accommodations. *Domecia acanthophora,* a little one-centimeter-long crab, lives within a chamber on elkhorn coral. Its relationship with the coral is more commensal (sharing) than parasitic. The crabs *Cryptochirus* and *Pseudocryptochirus* settle in pits on lettuce corals and the growing coral all but encloses them. Often, a male and female are found together. When abandoned, these mini-caves in the coral are quickly occupied by gobies.

Living associations within the reef are legion. Some are obvious, others more subtle. Between coral polyps, ciliated protozoans, flatworms, copepods, and other tiny crustaceans live and prosper no worse for wear either from the coral's feeding tentacles or mucous strands. Collectively, they look like pieces of sand on the polyps, quiet most of the time but occasionally making quick, erratic movements, then lying low again. Some can be found in the polyps, perhaps feeding on indigestible bits of food not yet expelled by the polyp. Many are species-specific: found nowhere else except within the confines of one kind of coral. To see them requires a low-power microscope. To separate them from the coral, immerse a small fragment of coral in seawater containing five percent alcohol, wait seven to fourteen hours, then collect and examine the sediment that has fallen off the coral. Rubbing alcohol will do; so will gin or vodka, which is sometimes more readily available. Remember that the proof is twice the percentage of the alcohol content, so make the concoction about one part gin to nine parts seawater.

The free-swimming **bristleworm**, *Hermodice carunculata*, larger than the northern clam worm but structurally much the same, feeds voraciously on **clubbed finger coral**, *Porites porites*, and staghorn coral. Be careful handling this worm. Its *setae*, the hairs along the sides of its body, stick into flesh easily and are irritating and painful.

Hurricanes

Physical damage to a reef, although sporadic, can be catastrophic. West Indian hurricanes strike with a fury few people comprehend. Crashing seas loosen and topple both elkhorn coral and brain corals, sending them crashing down on more fragile fore-reef species, burying them and hardier species in their path. The great clouds of sediment stirred up by the violent wave action resettle and choke out those corals not physically destroyed, adding significantly to the carnage. A hurricane may only strike a reef once in fifty years, but it may take half that long to repair the damage one leaves behind.

In 1961, a hurricane off Belize decimated its barrier reef. Regrowth took nearly twenty years to cover the debris left by that great storm. The north shore of Jamaica was severely battered by Hurricane Carmen in the autumn of 1974, by Hurricane Allen in August 1980, and again by Hurricane Gilbert in 1988. Extensive damage was done to corals as deep as forty-five meters, and the surf zone turned into piles of broken rubble.

Both physical violence and biological degradation are essential in reef-building, for the reef is a cemented conglomerate of large chunks and small particles. Some of the finer particles are carried away by heavy wave action, but enough rubble and sediment remain to build up the bottom at rates varying between 0.5 to 5 meters every thousand years. The pace of the fastest-growing reefs in the West Indies is fifteen meters per thousand years.

Kinds of Corals

Over sixty species of corals grow in the West Indies. In quiet waters, *Oculina* species grow from a single stem into a branched bushlike shape. *Favia fragum* form small, encrusting, cobbled knobs. *Astrangia solitaria* occurs in single cups, *Cladocora arbuscula* in tubed, irregular clumps, and *Madracis mirabilis* in tangles of tightly grouped branches in which small invertebrates hide.

But the corals we are most familiar with form massive stands or single structures. The **brain coral,** *Diploria labyrinthiformis,* forms giant boulders up to three meters in diameter. *Montastrea annularis* grows in enormous vertical and horizontal sheets, often forming horizontal plates layered in tiers. It cloaks the buttress walls of old coral stacks in knobby, irregular, and discontinuous growth that looks like so many errant roof tiles.

In between and among both the hard and soft corals there grows a false coral, called **fire coral**, *Millepora alcicornis*. It comes in a bewildering array of shapes but all are a light mustard brown often tipped in white. Beware! On tender portions of human skin it packs the wallop of a hot soldering iron and raises a welt that lasts for weeks. It is not a true coral, but a hydrozoan. Although it can encrust anything firm, it also can be found free-standing in flattened or folded sheets, like drapes or wrinkled curtains that rise vertically from a lumpy, encrusting base.

Common corals can be identified by sight but some lookalikes must be sorted out by a few simple measurements and a closer look. When coral-watching, carry a short plastic metric ruler with you. Measure the average diameter of the cup in which the polyp sits. If the cups join together to form valleys, as they do in brain corals, measure the width from wall to wall. Partitions radiate from the walls of the cup into its interior and, occasionally, out beyond its periphery. The numbers and distance between these partitions also aid in identification, as does noting the absence or presence of a center structure, the *columella,* in the cup and its shape.

Cross section of brain coral, *Diploria,* shows the progression of
outward growth of individual polyps.

Fire coral, *Millepora*

If you have access to a rock saw with a water-cooled carbide blade, cross-section a piece of dead brain coral and see for yourself how the polyps have built cup upon cup, outward radially. Don't pick live coral. It is illegal in Florida, will impoverish the reef, and the coral serves very little purpose when dead.

The undersides of loose dead coral are worth a look; aside from the vacant homes of tube worms and boring clams, they often contain small, bright red, knobby protuberances of the foraminiferan *Homotrema rubrum*. This shelled protozoan can be so abundant that it colors the coral sands a light pink, as it has done on the beaches of Bermuda.

Tropical coral polyps close during the day. At night they open in full display, combing the waters for a tiny meal. Look closely at their tentacles, especially those of *Montastrea cavernosa*, whose cups are nearly one centimeter (nearly half an inch) in diameter. All its tenta-

Close-up view of the open polyps of the coral *Montastrea cavernosa* (Photo by Robert Bachand)

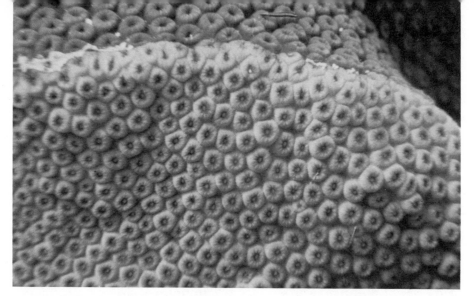

Where corals meet, there is a struggle for dominance. *Montastrea cavernosa, top,* and *M. annularis, bottom.* Note the dead zone in between.

cles are about the same in length save one, the sweeper tentacle, which is ten times longer than its neighbors. This tentacle can reach over several polyps and will attack alien coral encroaching on its borders.

As two corals of different species grow toward each other, there comes a time when they compete for the same space. You might expect the faster-growing species to win out by simply overwhelming the other, but the slow-grower can alter the odds by fighting back. When *M. annularis* and *M. cavernosa* meet, the sweeper tentacles of *M. cavernosa* attack and kill the polyps of *M. annularis,* leaving a no man's land between, where nothing grows.

The skeleton of a single cup of the star coral, *Montastrea cavernosa*

Those that have no sweeper tentacles do battle with the long, threadlike digestive filaments they extend outside the polyp into the polyp of the enemy. A hierarchy of winners and losers—the smaller flower corals among the winners and the club finger corals with the losers—has been established by tank experiments. However, the pecking order in the laboratory has not always corresponded with observations in the field.

REEF FISHES

All the spaces in a coral reef make marvelous havens for hundreds of species of reef fish, who range from thumbnail size to more than a meter long. A busy reef will have a cloud of hundreds of small fishes swimming over it. The value of the nooks and crannies below them becomes quite apparent when a predator approaches. The cloud of fish shrinks as the hunter cruises in and, if frightened by sudden movement, vanishes instantly into the protective coral cavities. As the threat passes, the little fish quickly emerge and go about business as usual.

Shapes, colors, and markings, of reef fish seem without limit. This smooth trunkfish is encased in a triangular box of bone.

A queen parrotfish being cleaned by yellowhead wrasse.

Identifying all these fish is far easier than recognizing cold-water species. Their shapes, colors, and color patterns are distinctive enough to remember while thumbing through a guidebook after a dive. The common ones will be seen on every dive, and every visit to the reef will add more to your species list. A week of visits should net over fifty species positively identified and dozens more tentatively so.

But if you have an initially high success rate, don't be lulled into careless observation. Among some fish, color patterns are quite variable. In some closely related species, color patterns and other differences can be slight. Note color, size, shape, and specific markings. Does the fish have a distinct spot? Is it all one color or multicolored? If it is striped, how many are there and what is the color sequence of the stripes? Is it barred (vertical stripes)? Is the body slender, exceptionally thin, unusually long, boxlike? Are the fins unusual: prominent spines, thread-finned, fanlike, or stalked? Where did you see the fish? Very close to the surface, midwater, or on the bottom? Sitting on the bottom, in a burrow or a coral crevice? How big is the fish, a half-centimeter or one meter? How does it behave? Does it school, mix in with others, or is it a lone swimmer? Does it lay still, walk on the bottom with its pectoral fins, wriggle into the loose bottom sediments?

French grunt in John Pennekamp Coral Reef Preserve, Florida

Is it associated with an invertebrate?

Look and read, then look and read again. Within a few dives you will recognize the differences between **French grunt,** *Haemulon fla-volineatum,* and the similarly colored **porkfish,** *Anisotremus virginicus,* who has two black vertical bars on its head. Keep the Latin names in your notes, for if you decide to find out more about them, the Latin name is the key to the scientific literature.

After a few tropic trips, you will become aware of the typical body shapes of the common fish families; that is, you will recognize a parrotfish from a snapper or a grouper even though you do not recognize the exact species. There really isn't any reason to try and identify every fish you see. What is more important is the variety of ways each makes its living and how it interacts with others in its world. If that interests you, your lifetime isn't long enough to take in the half of it.

The behavior of reef fishes is as intriguing as their splashy coloration and bizarre shapes. Many divers are in such a hurry to see another part of the reef that they move on long before they have had a chance to absorb the rhythm of the life before their eyes. As Yogi Berra once said "You can see a lot just by lookin'," and if you settle down and watch, you will.

Some fish stake a claim to specific territory and defend it with a belligerence that belies their small size. For the damselfishes, no chal-

Algae can quickly overrun
a reef if their predators
disappear or if the waters
are polluted with sewage.

A French angelfish swims over the star coral
Montastrea annularis

Condylactis gigantea, largest of the common Caribbean anemones. It is often the home of the small spotted cleaner shrimp.

lenger seems too large. Algae-eaters, they will vigorously defend areas of dead coral on which algae rapidly proliferate. Charging, retreating, then charging again at all comers, large grazing fish—tangs, surgeonfishes, and parrotfishes—retreat after barely a brief nibble.

Some damselfishes are said to bite off coral in order to "farm" the ensuing growth on the broken surface. Most are plankton feeders. They snatch copepods from the water with a curious snap of the head. When feeding on plankton, the damsels suffer a curious angst: Pickings are better the farther away they get from the reef, but their uneasiness grows rapidly with the distance from home.

Other territorial fish patrol a set pathway in the reef, weaving in and out of crevices and only abandoning the outside leg of the circuit when faced by imminent threat. Some hover in reef recesses and simply withdraw deeper if intimidated. Many are awaiting the night shift. Squirrelfishes feed at night, venturing away from the reef only under cover of darkness. Day-feeders replace them, resting in the very same holes vacated by the night foragers.

So many reef species share the same digs and appear to make a living in similar ways that biologists have been puzzled by this seeming contradiction of the ecologist's dictum that says no two species will occupy the same niche at the same time for too long; one of the competing species will gain a competitive edge and displace the other. How then does the reef support hundreds of species with such similar lifestyles without the rise of a few dominant species? Some experts claim each species is specialized enough to exert a subtle claim to its own turf. Others argue that all are so well-suited to their environment that space and successful existence is allotted by an "endless lottery" determined by chance.

Mimicry and subterfuge are common ploys among reef fish. Less aggressive fish cloaked in the colors of a more aggressive species gain an important margin of safety as a predator hesitates, unsure of his victim. This gives his intended prey an opportunity to widen the striking distance beyond the predator's rushing range. Such disguise can be as simple as spots on the flanks. Members of the butterfly fishes have spots that may appear like large eyes to a following fish. Is the predator really put off by this elementary ruse, or is this explanation of the purpose of the spots the figment of some biologist's imagination? It is hard to say. Fish obviously don't like approaching "eyes," and will back off—that is easy to observe on nearly every dive—but whether this observed behavior specifically protects those butterfly fishes with spots may be an unwarranted assumption.

And the game cuts both ways. An aggressor who mimics a benign species, either by taking on its color patterns or imitating its behavior, can narrow the gap between itself and its intended victim.

The clever use of cover also gets results. The **trumpetfish,** *Aulostomus maculatus,* a long, slender fish-eater, will swim alongside algae grazers such as parrotfish, using them as a blind to get within striking distance of small wrasse, who are unperturbed by the parrotfish's approach. It will also hover vertically, head down, aligned with a sea fan, and wait for passing prey. The trumpetfish's strike is short and sure, and so fast your eyes cannot follow it. One second earlier the little fish is there, and in the next instant, gone.

Not all relationships are so ruthless; many are distinctly cooperative. Cleaning, one species picking parasites or dead matter off another, exists between fish and fish, shrimp and fish, and crabs and fish. Most of the cleaner-fish are small gobies, blennies, and wrasses. The **neon goby,** *Gobiosoma oceanops,* electric blue with black stripes, the **cleaner goby,** *G. genie,* and the juvenile **bluehead wrasse,** *Thalassoma bifasciatum,* congregate at work stations and signal their services by an undulating display dance, weaving back and forth over their favorite coral head, awaiting customers. The approaching host must also act its part by flaring its pectoral fins, aligning its body head-up or -down, or holding its body diagonally.

The cleaner will then approach the host, lightly butting its flanks. The host responds by holding still in whatever odd position it had earlier assumed. **Creole wrasse,** *Clepticus parrai,* align head-down, as does the **queen parrotfish,** *Scarus vetula.* Other parrotfish angle their bodies head-up, while the **doctor fish,** *Acanthurus chirurgus,* turns sideways, not quite horizontally but close to it. Just how variable or universal these postures are within a given species is an open question. Once you get to know these fish, keep notes of your observations at different cleaning stations. You may be able to shed some light on this behavior pattern.

The cleaners work on one fish at a time, sometimes concentrating on mouth and gills, sometimes on flanks and tail. When the host has had enough, a few shakes of its body sends the cleaners away, who wait for the next customer to move up in the queue. As each fish awaits its turn, now and then a cleaner will break away from the main group and briefly butt its flanks as if to assure it that it is next in line.

Every reef of reasonable size has a cleaner station centrally located, but not necessarily at its crest. To find it, simply watch the parade of passing foragers. Should a few stop their swim-and-peck routine and assume an awkward stance, you will soon see the cleaners appear. You can approach the group, but the cleaners may get nervous and leave if you come too close.

Small fish often follow large grazing fish for short distances, gobbling up whatever invertebrates are uncovered by their foragings. Uprooted and momentarily out in the open, they are easy pickings for the wrasses and blennies who tag along until the larger fish move too far into unfamiliar territory. For the wrasses, gobies, blennies, and even butterfly fish, getting too far away from home may mean not being able to navigate back.

On home ground, small fish behave more opportunistically than do larger ones. Break open an urchin and watch the local juveniles arrive first. The bluehead wrasse, **slippery dick,** *Halichoeres bivittatus,* and other small residents are the first on the scene, grabbing what they can before bigger fish butt in and hog the meal.

Even large reef fish keep to a limited home range. Tagged groupers and snappers have been recaptured where they were first caught as much as three years later. Parrotfishes return to their favorite caves to spend the night, but may roam considerable distances during the day.

THE REEF AT NIGHT

At night, the cryptic life of the reef emerges. Shrimp and crab venture out to feed. Urchins leave their crevices to graze on algae in nearby sandy bottoms. The basket stars, whose days are spent balled up in a crevice or wrapped around a sea fan, unwind, spread their tentacles out, and hunt for plankton. Brittle stars also seek the passing plankton by raising three of their five arms.

The moray eels go hunting. The **spiny lobster,** *Panulirus argus,* emerges, as does the **common Atlantic octopus,** *Octopus vulgaris.* If you come across an octopus at night, you will find it totally bedazzled by your dive light. You can pick it up (gently) and it will offer little or no resistance and will not attempt to bite you. If you play too rough, it will discharge a cloud of ink and flee, a not too unexpected reaction. During the day you can tell if its den is close by if you come across a grouping of clean conch shells. Conch is a favorite food of the octopus, and invariably it brings the conch to its doorstep before eating it.

Some fish change color at night. The **spotfin butterfly** fish, *Chaetodon ocellatus,* develops a dark, smeary blotch on its white flanks. The color of the **blue tang,** *Acanthurus coeruleus,* a deep blue by day, fades at night, and four vertical gray bars appear on its side. The bright red **cardinal fish,** *Apogon planifrons,* turns to a dull pink.

Divers' lights at night alter fish behavior. The **spotted goatfish,** *Pseudupeneus maculatus,* has three or four dark brown blotches down its flanks by day, but variegated red blotches at night. Or are the night colors caused by a fright reaction to the blinding lights of the diver? Stress induced by light causes red blotches in the **Pacific goatfish,** *Mulloidichthys dentatus.* Is the same true for its Caribbean relative?

Lights in or on the water at night attract hordes of plankton, who are quickly followed by **bigeye scad,** *Selar crumenophthalmus,* who home in on the life dancing in the beam. Needlefish and herrings also join the fray. If the beam is on long enough, squid will dart in for a meal, as will an occasional barracuda.

Night dives are best made from an anchored boat. If it is your first night trip, buddy up with someone you know and trust. Don't stray too far from the boat. It isn't that the blackness holds long-leggity beasties, or ghosties and ghoulies, but the fear of what you can't see may induce a claustrophobic response. If apprehension *does* sweep over you, don't fight it. Go back to the boat and await another time. Try again, and next time start in more benign surroundings. Once you are caught up by it, the beauty of the reef at night will draw you back just as its beauty by day will entice you to it whenever you have the opportunity to return.

EXPLORING THE REEF

A word of caution when exploring the breaker zone of the coral reef: The inner fringes of elkhorn coral create a maneuverability problem for the diver. An errant surge of water can hurl you into the crest of the reef with unanticipated suddenness.

Never swim over the top of a shoal reef. Doing so is a prelude to grief. If you are dropped into the trough of a passing wave, the power of the following sea will push you farther aground, scraping your hull from stem to flipper. Deep coral cuts and abrasions head to foot will ruin your whole day, to say nothing of the damage you may do to the reef. And if fire coral is part of the fauna in the breaker zone, you may spend the next few days in a hospital ward.

Skirt shallow zones carefully, penetrating inward only where a wide stretch of open deep water abuts the shoals so that you have plenty of time and water in which to execute a tuck turn. If carried by a wave, meet the shallows fins-first, then beat to sea as the wave passes and its energy slackens. At worst, you will only suffer scratched ankles.

Unfortunately, many reefs are not what they were a mere forty years ago. Silt drifting seaward from construction projects, sediment stirred up by boats, overnutrification from sewage—in short, the depredations of man—have left the waters cloudier and the reefs poorer every year. You can help preserve what remains: Take only pictures, don't let your fins stir up sediments or break off coral, and don't let your boat anchor scrape over the reef.

Plan in advance how you intend to get to the reef and safely back. Charter trips are available throughout the Keys and the Caribbean.

Should you bring your own boat or rent a bare boat, consider three important elements to a successful day of exploration: the weather, the sun, and finding your way home. Strong winds off the Atlantic can kick up a wicked chop in reef-filled shallows, making it unsafe to navigate among coral heads, yet difficult to remain anchored.

For the untanned, the sun's intensity, even in winter, is a serious threat. A day of overexposure can mean a week's incapacitation. To solve the problem and at the same time protect your skin from coral abrasions, consider wearing a lightweight full-sleeved shirt and pants over your swim suit. Add a pair of light cotton gloves to protect your hands. You won't look the height of fashion, but at the end of the day you will be in one piece, unburned and unscraped.

Then there is the problem of finding your way back to the dock. Obviously, in the Keys, west is home, but finding a channel into a mangrove with the setting sun in your face is not often easy. Off Florida, the mainland Keys are low with few prominent landmarks, so keep track of every useful bearing you can. A chart is a must and a VHF radio a wise complement. These waters look benign and inviting, but have created their share of grief for sailors over the years.

Coral reefs contain so much exhilarating life that the urge to hold in your memories is bound to lead you to underwater photography. No other place holds so many photogenic possibilities, and no other way allows you to harvest its bounty so easily without doing harm. Those photographs and your log, years hence and perhaps in less propitious circumstances, will remind you of blissful times past and the beauty you were once privileged to see.

8

Tropic Lagoons and Mangroves

BEHIND THE REEFS AND IN THE QUIET WATERS OF LEEWARD shores lie submerged flats of coral sand and grass. In shoal waters these may meld into mud flats and stands of mangrove where the difference between land and sea becomes hard to discern. Each domain claims its own unique inhabitants while sharing others among all. These communities are often overlooked by the snorkler in favor of the more exotic visual pleasures of the reef, but the backwaters hold as many surprises and curiosities as do their more flamboyant neighbors.

SAND FLATS

The incessant creation and destruction of carbonate-bearing plants and animals has, over eons, created bottoms of limey mud, hardened marl, and loose sand. Sand flats line the west coast of Florida from the Keys to the Panhandle, inshore and offshore well into the Gulf of Mexico. Sandy bottoms also lie in the lee of windward reefs and promontories: pulverized reef particles washed inshore. Some sand flats run unbroken with little relief over wide expanses, while others are pockmarked by bare, hard bottoms, isolated patch reefs, mud banks, and underwater stands of grass.

If a small clam, *Codakia costata,* about one centimeter long, grows in abundance on a sand shoal, the location is sure to be locally heralded as a bonefish flat, for the clam is that fish's favorite food. Schools of **bonefish,** *Albula vulpes,* enter the shallows on a rising tide, rooting for clams with their blunt snouts. You can see their tails protrude from the water as they blow and grub in the shallows for a meal. Once a clam is engulfed, the bonefish grinds it up between patches of rasped teeth set in its upper jaw and on the upper side of its tongue.

The **ten pounder** or **ladyfish,** *Elops saurus,* also feeds on the flats with the tide and can often be seen skipping along the water's sur-face—just why they do this is a mystery. Drums, the sea trouts, *Cynoscion* species, and **red drum,** *Sciaenops ocellata,* also prefer sandy shallows. Over thirty species of gamefish swim these flats; some move out to deeper channels to hunt, others move in from brackish water to forage. Some are motivated to travel by sudden variations in water temperature and salinity. Florida Bay water temperatures can plum-met during occasional winter cold snaps. Salinity in the Bay is not only lowered immediately by rainfall, but changed seasonally by vari-ations in runoff from the Everglades.

A few small reef fish, generally wrasses, make homes among the dead patches of coral rubble and outcrops that pockmark the flats. These inner-reef fragments contain few live corals and are often overgrown with algae and riddled by boring organisms. They are reg-ularly visited by parrotfish, who bite off hunks of coral to get at the algae growing over the surfaces. As you swim underwater over sand, you can often hear the light scratching sound of a grazing parrotfish before you see the patch reef. Water clarity inshore is seldom as good as that over the fore-reef farther out to sea. Wind chop and wave action in shallow water quickly resuspend the fine bottom sediments.

Water clarity also worsens as the day progresses. Inshore water over coral sand and patch reefs is saturated with calcium carbonate. Photosynthesis by algae shifts the acid-alkali balance in the water; the bicarbonate-carbonate ratio shifts such that carbonate begins to come out of solution. This turns the water just milky enough to enshroud the underwater scene in a light fog.

If a reef is nearby, groups of **spiny urchin,** *Diadema antillarum,* may be out grazing on algae growing on sand grains. Once prolific in the Caribbean, a disease carried by the currents has dropped the urchin's numbers precipitously. If you have ever blundered into one of these creatures and suffered the painful consequences of their sharp spines you may be inclined to say "good riddance," but their virtual disappearance from some places has led to unchecked algae growth that has smothered other sessile life.

While swimming over sandy bottoms, you are likely to encounter two mysteries: cones of sand and a fragile, gelatinous streamer shaped like a woman's nylon stocking emanating from a half-dollar-sized hole. The latter holds the eggs of a polychaete worm nearly twenty-five centimeters long, the **lugworm,** *Arenicola cristata,* who builds a U-shaped burrow in silty bottoms and casts out this casing after breeding.

The mounds may either be the work of a burrowing sea cucumber, *Holothuria arenicola,* or the ejections of a burrowing shrimp, one of the Callianassae, kin to the ghost crab common to southern beaches. If the cone erupts a shower of sand grains, it's a shrimp; if you find cylindrical sand casts on top of the hillock, it's *Holothuria,* a sea cucumber.

Sand bottom is a good place for an animal to hide and, therefore, a good place to hunt for a meal. A stingray can go virtually unnoticed when buried in the bottom. The lizard fishes, Synodontidae, blend in so well they have only to sit and wait for a morsel to pass by. Small groups of porgies grub in the sand for mollusks and crab. Goatfishes stir up the bottom with their chin barbels, uncovering worms.

The waters over sand abound with shoals of herrings, sardines, pilchards, and anchovies—all silver-sided sliver-shaped baitfish—that require close attention to tell one from another. The common **Atlantic thread herring,** *Opistonema oglinum,* has a long ray trailing from its dorsal fin that is often tipped in black, as are the ends of its tailfin, but the ray lies nearly flat against its back as it swims. The **red-eared herring,** *Harengula humeralis,* has an orange spot on its gill cover and streaks of orange along its sides. Other herrings are streaked in blue and emerald green and, except for differences in body shape, look much like the silversides with whom they intermingle.

Snorklers and divers invariably look down, rarely ahead and up at the undersurface of the waves. That is where you will see the needlefishes and halfbeaks: slender, silver-green spears in schools swimming barely submerged. They will leap out of the water at the slightest provocation: to pass a floating twig or to avoid a potential enemy.

Learning how to pick up and hold this
formidable pin cushion, *Diadema
antillarum,* on a single breath of air is a
sure sign that your underwater skills
are well above average.

Egg case of the lugworm, *Arenicola*

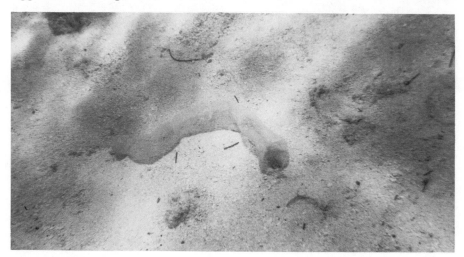

Sand mounds in turtle grass made by a burrowing shrimp. Every so often these miniature mountains erupt a shower of sand grains.

The **houndfish,** *Tylosurus crocodilus,* nearly a meter long, leaps so violently as to be positively threatening to someone sitting in an open skiff at night. Alarmed by a sudden sound or flash of light, this "living javelin" can hurtle itself into the air in an arc that one would do well to avoid, even though the distinction of having been speared by one would provide you with a corker of a fish story.

The occasional hard outcrops in sand flats, be they old coral mounds or exposed ledges or overhangs, are often riddled with small caves at the sand line. This is where the spiny lobster makes its home, sometimes alone and sometimes sharing its digs with others. If one antenna faces forward and the other points backward, it is a sure sign the lobster is keeping track of another occupant—a moray eel, for example.

The spiny lobster is relatively large. You may come across a smaller, more colorfully marked relative, *P. guttatus.* There is a third common species, the **slipper lobster,** *Scyllarides aequinoctialis,* which blends so well into its surroundings that divers usually overlook it. Its antennae have modified into paddles. Its body is a series of flat, armored plates that gives it the appearance of a giant pill bug. The reason it's so easily missed is that it clings to the roofs and walls of coral caverns, not to their floors.

GRASS FLATS

Beds of grass flourish in quiet and shallow waters between the edge of the beach and the crest of the coral reef, in bays, and in waters adjacent to mud flats and mangrove stands. These grasses, predominantly **turtle grass** *Thalassia testudinum,* and, less often, **manatee grass,** *Syringodium filiforme,* and **shoal grass,** *Halodule wrightii,* never emerge from the water nor grow where the bottom is laid bare at low tide.

Like its northern cousin, *Zostera marina* (**eelgrass**), turtle grass is broad-bladed (1.2 centimeters wide) and up to a third of a meter long. The blades rise from dense networks of runner roots well set beneath the sand and sediment, which they securely bind. Along tropic shores, turtle grass and mangrove leaves supply more bulk biomass to the waters than does the primary productivity of phytoplankton.

Most of the turtle grass reproduces vegetatively, but each year a certain fraction of it flowers and produces seed. These diminutive three-petaled white blossoms bloom in spring and summer and can be seen close to the root stems.

Turtle grass can grow in mud, sand, and broken bottom, from shallow water to depths of fifteen meters, and in waters whose salinity varies from twenty-five to thirty-eight parts per thousand. It does best in temperatures between 20° and 30° C. Above and below those temperature extremes, its leaves die off, but they will regrow readily as temperatures return to the preferred range.

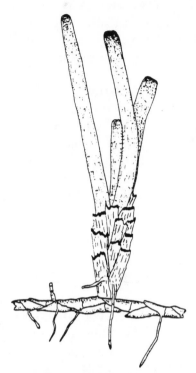
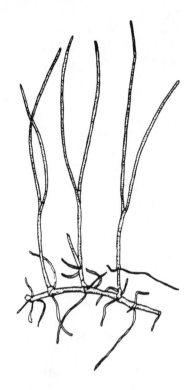

Turtle grass, *left;* manatee grass, *right*
(Illustrations by J. M. Foster)

In Florida, Turtle grass may intermingle with manatee grass. Unlike turtle grass, the blades of manatee grass are round in cross section and much narrower (0.3 centimeter). Both may commingle with shoal grass, which looks like a smaller version of turtle grass. Its leaves are flat but only a fifth the width of *Thalassia.*

Turtle grass turf creates living quarters and hiding places for a multitude of plants and animals, large and small. Over 113 species of tiny algae live on its leaves. Almost all are *epiphytic,* that is, they use the leaves just for physical support and take no nutrients from their host.

Poking about in the grass will uncover the large solitary algae: the fan-shaped *Udotea flabellum,* the thick, flat, spongy blades of *Avrainvillea nigricans,* and the **mermaid's shaving brush**, *Penicillus* species. The latter is bright green, while the former two are very pale green. (A recent manual has substituted "Merman" for "Mermaid," no doubt to correct the somewhat inaccurate gender reference.)

Mats of *Halimeda,* a multiple-stalked coralline alga, often fill open patches in the grasses. Each stalk holds tiny tri-lobed calcified disks strung along its stems like so many green buttons. Fan worms, among others, set up housekeeping between the disks. *Halimeda* is the source of much of the calcareous sand found in reef lagoons. Browsing fish gobble it up whole and pass out its calcified remains as a fine sediment. This sediment settles between larger grains, into cracks and crevices of larger fragments of coral, and eventually binds all into a hard marl that, over eons, consolidates into limestone.

Animal and plant associations are commonplace within the grasses. Bring along a zip-lock bag to discover who lives with whom and how. Gently tear a specimen loose from its mooring and trap all you can in the plastic bag. Keep it cool until you can transfer its contents to a flat dish. Spread out the material and see how many species you uncover. Sketch or photograph what you see and preserve the specimens for later identification. You will be overwhelmed by all the small forms new to you.

The silt in *Thalassia* beds is so fine that living beneath the surface is restricted to animals that can tolerate low oxygen. Polychaete worms do well: That environment is their specialty. Small clams of the family Lucinidae burrow deeply here, but other mollusks can go down no farther than they can extend their siphons, or, like *Pinna carnea,* the **pen shell,** they expose the topmost portion of their shells while burying the remainder into the sediment.

Foraminifera, one-celled creatures with calcified shells that are so small they usually require a microscope to be seen, have two giant cousins on the grasses that can be seen with the naked eye. Look for very small coiled shells and round buttonlike shells on the blades of the turtle grass. The former is snaillike **turtle grass foram,** *Archaias angulatus,* and the latter the **button foram,** *A. compressus.* Unlike snails, the growth lines on these shells run *with* the curvature of the shell, not across it. Forams have tiny pores in their shells from which they extend their cell walls for feeding. If you have a microscope, collect a few and check this out for yourself.

Turtle grass is also home to a few species of solitary corals. They usually grow in thin spots or along the edges of the grass bed where currents keep them relatively silt-free. **Rose coral,** *Manicina areolata,* is the most common. A flattened yellow-brown oval with undulated edges, seven to ten centimeters in diameter, it lies unattached on the sand. Patches of tan fungus-shaped *Agaricia agaricites,* ochre knobs of

Favia fragum, small crusts of *Siderastrea siderea,* and pebbles of *S. radians* show up here and there but are not abundant. These small outcrops usually attach onto limey fragments tossed into the lagoon by past storms.

In deeper water that is free from sediment, the delicate **ivory bush coral,** *Oculina diffusa,* grows in branched clumps. Its larger relative, *O. valenciennesi,* has a main stem as thick as a thumb and more open branches. Both need a firm base from which to grow to full size.

Feeding on organic matter in the silt, the light sienna to dark brown sea cucumber *Holothuria mexicana* is common in Caribbean grass flats but not in Florida. Sea cucumbers are echinoderms, relatives of the seastars. Except for the rows of tube feet on their undersides, their leathery skin and sausage shape doesn't hold much likeness to either the starfish or the sea urchin. Their body plan looks simple enough from the outside: a tentacled head at one end and an anus at the other. Inside, however, they have a complex water vascular system that, through a ring canal, works the tube feet and the tentacles. The gut has three large loops. Without a circulatory system, digested food either diffuses to where it's needed or is carried there by wandering amoeboid cells. Sea cucumbers exchange oxygen from the water with a respiratory tree near the anus that draws in water at that end, extracts the oxygen, then expels the water. Slow movers, their tracks are marked by cylindrical fecal pellets made up almost entirely of sand and silt.

Actinopyga agassizii, a common sea cucumber, is often home for the **pearlfish,** *Carapus,* which lives within its anus. To pop it out, put the sea cucumber in a bucket of seawater and give it a shake, but don't overdo it. Sea cucumbers protect themselves from predators by ejecting their innards or shooting out a sticky excretion. Either defense will leave you with a very messy bucket.

Two sea urchins are common in turtle grass: the green-and-white *Lytechinus variegatus,* about seven centimeters in diameter, which bears short reddish brown spines, and *Tripneustes ventricosus,* the **sea egg,** twice the diameter of *Lytechinus.* The test of *Tripneustes* is purple or dark brown, and its short spines are white. Both urchins cover themselves with bits of grass, coral, pebbles, and broken shell. Neither feeds directly on live grass, but on dead debris. Most permanent grass inhabitants do not eat green grass; it's the occasional visitors like turtles who directly feed on it.

Several species of sea hare, *Aplysia,* large shell-less gastropods, frequent the grasses. One is spotted with bulls-eye rings, another is

The sea egg, *Tripneustes ventricosus,* covers itself with bits of shell, grass, and whatever else is loose and handy.

green, another black, and one is so ragged it appears to have been on the losing side of a battle. All exude a beautiful purple ink when squeezed. Every diver who sees this for the first time is tempted to give it a try. If the location has but a few sea hare and many divers, it must come as a welcome relief to the sea hares when the divers depart.

The sea hare, *Aplysia,* browses on algae

The West Indian sea star, *Oreaster reticulatus*, is now scarce where tourists are abundant. (*Photo by John Storr*)

Few seastars make their home in turtle grass, but one of them, the huge **West Indian star,** *Oreaster reticulatus,* makes up in size for its rarity. A brilliant orange-red, it is found only in turtle grass. It feeds on sponges and small sea urchin. Because of its size (the length from the tip of one arm to another across its center can approach two-thirds of a meter) and its bright color, it has been overcollected and is now rare.

So too are the conch and helmet shells, collected both for food and the curio trade. The **queen conch,** *Strombus gigas,* has a twenty-five-to-thirty-centimeter-long shell and a living weight of two kilos. A herbivore, it uses its long proboscis to scrape algae off turtle grass blades and stony debris. Its forward motion is a wonder to behold. It extends its muscular foot, digs in the horny operculum at the end of the foot, then violently contracts, jerking its body forward in more of a

A collection of conch shells, freshly emptied, found near an octopus lair

lurch than a walk. If you pick it up, the foot retreats into the shell and a horny door, the operculum, perfectly blocks the entrance. As you continue to hold it, two baleful eye stalks will emerge to survey its dilemma. Slowly, the foot will move out over your hand, then suddenly contract in an effort to dislodge you. Let it go unless you know how to make a meal of it properly and have a liking for its meat. The shells of the dead are too frequently all you will ever see of this magnificent creature.

As the queen conch ages, its shell is attacked by all sorts of boring organisms. The shell thickens by adding material to the interior to patch the excavations of its enemies. As time passes, the shell gets heavier and the animal inside lighter and smaller. The knobs that line the spire erode away. These old conch were, at one time, thought to be a separate species and were called "samba conch."

Conch harbor houseguests. The **conchfish,** *Apogon stellatus,* five centimeters long and a mottled silvery brown with large pelvic fins, stays inside the conch by day and feeds outside at night. An empty tin can will serve the conchfish as well. If you find an empty soda can on the bottom, not a rarity these days, pick it up and give it a shake inside a plastic or mesh bag. If you are rewarded with a conchfish, well and good—just remember to take the can back to a trash bin when the day is over.

The **fighting conch,** *S. pugilis,* is similar in appearance to the queen conch but smaller, only ten centimeters in length. Its shell interior is salmon-colored, not pink as is the queen conch. You will rarely find an immature queen conch during the day. Juveniles bury themselves in the sand, emerging at night to feed. The only time an adult queen conch burrows in sand is to lay eggs—a long string neatly bundled in a packet that, if unwound, is twenty meters long and contains a half-million embryos.

Humans are the main enemy of the conch. The West Indian fishery catches them by the millions and every settlement near a productive grass bed has its huge shell heap. Tiger shark have been known to swallow conch whole, and loggerhead turtles eat them by crushing their shells. The **eagle ray,** *Aetobatus narinari,* is said to eat juvenile conch. Its digestive system apparently dissolves their shells. The stomach contents of one ray turned up forty-one conch without any sign of shell or operculum! Octopuses also feed on these and other large snails. Invariably, they bring the snail back to their lairs to feed. When finished, the shell is pushed outside. If you spot several newly emptied shells on the sand adjacent to a patch of limestone reef with holes or ledges, you can be sure the home of an octopus lies within.

Few sea creatures hold more fascination than the octopus. A master of camouflage, his mottled and varigated skin colors match the background exquisitely, obscuring his shape from the casual eye. Discovered, he will flush in embarrassment, altering between light and dark hues. Flight follows fright; an inky cloud marks the decision. Pursue him, if you can, and see what happens next. If home base is close-by, he will hole up. If not, he may duck into any convenient orifice, often startling another inhabitant who is unprepared for his company. Back off a bit and await developments. Perhaps a spiny lobster or moray eel will appear, displaced by the agitated intruder. Wait long enough and the octopus will peer out to see if the coast is clear. As you retreat, he will tentatively emerge; if you advance, he

Caught out in the open, the octopus will quickly change color and flee, often ejecting a cloud of ink to mark its departure.

usually opts to stay where he is. You may find yourself in a game of peek-a-boo with this bright sharp-eyed creature, amusing to you but very serious business to him. (The gender of our new acquaintance may just as easily be female as male; but once personally encountered, we can hardly refer to him as "it.")

Octopuses feed at night. In the beam of a diver's light they appear totally bewildered and can be handled (gently) with impunity until the impulse to flee carries them beyond the glare of the offending beam.

Five species of octopus are common on reefs and grasses: the largest, *Octopus vulgaris,* a meter long from head to tentacle tip, is the most common.

MANGROVE

Thick tangles of mangrove line long expanses of low-lying tropical coasts. Growing in the quiet shallows, they stabilize the fragile shoreline against storm damage and erosion. Their exposed roots harbor hundreds of kinds of animals; their fallen leaves feed countless more.

Stands of mangrove are so thick and interwoven as to be impenetrable from landward. The best way to approach them is to wade or swim up to them from a nearby beach. Don't be put off by their mild "rotten egg" aroma.

Closest to the sea, the **red mangrove,** *Rhizophora mangle,* is successively replaced further inland by the **black mangrove,** *Avicennia germinans,* **white mangrove,** *Laguncularia racemosa,* and **buttonwood,** *Conocarpus erectus.* The order of succession depends on the contour of the high ground and the saltiness of the water. White mangrove does better in brackish water than does red mangrove, and often lines the water's edge in less salty lagoons. Black mangrove will grow on land, but stands do well in salty tide ponds. Buttonwood does best farther inland. All mangroves need salty soil.

If you intend to explore around mangroves, learn to recognize **manchineel,** *Hippomane mancinella,* sometimes called *poisonwood* or *poison apple.* It looks much like an apple tree, with shiny dark-green leaves and green fruit. The milky sap in its branches, trunk, leaves, and fruit is intensely irritating to skin and mucous linings. Raw sap will cause blistering, swelling, and pain. You can be affected just by standing beneath the tree in a rainstorm. Manchineel has been eradicated in Florida but is common throughout the West Indies.

The prop roots of the red mangrove grow in looping tangles. Long cigar-shaped pods hang from its branches. Ripe pods fall and stick base-down in the shallows beneath the tree or float to new locations and take root as much as a year after entering the water.

Black mangrove roots run out radially from the main trunk and send spiky quills (*pneumatophores*) upward above the waterline. It can withstand the much saltier water of tide ponds than can either red or white mangrove. These trees trap sediments that maintain and build shoreline. Their decaying leaves produce immense loads of fodder that enters the food web in surrounding lagoons and flats. Up to sixty percent of the basic foodstuff supporting tropical estuarine life originates here; the rest comes from sea grasses.

Red mangrove invading new territory. As this clump of trees multiplies and grows, the surrounding sediments will increase, stabilizing the shoreline.

Mangrove root tangles. The waters around them are second to none in productivity and nurture the young of many different species.

The Florida Everglades contain the largest stand of mangroves in the United States, supporting vast numbers of terrestrial and aquatic animals. Extensive mangroves in the Keys are nursery grounds for shrimp. In turn, this nursery nurtures and feeds the **gray snapper,** *Lutjanus griseus,* Mullets, *Mugil* sp., **snook,** *Centropomus undecimalis,* and **tarpon,** *Megalops atlantica.* Tarpon can attain a length over two meters long. They live in the dark oxygen-poor mangrove shallows, gulping air by rolling at the surface.

Red mangrove support a rich mix of life above and below the waterline. Submerged stilt roots provide attachment sites for a dozen common algae: Little bushy fronds of *Caulerpa racemosa* and *C. mexicana;* horn-shaped buttons of *Acetabularia crenata* and green grape clusters of *Bostrychia* grow wherever enough light allows. If the suspended sediment is not too thick, the algae compete for space with the sponges: the fire-red *Tedana ignis* (careful, it's a finger burner), red *Haliclona rubens,* the branching yellow *Verongia,* green crusts of *Adocia carbonaria,* and massive gray lumps of *Ircinia strobilina* that house myriad crabs and brittle stars.

You will see tunicates on the roots: orange vases of *Ecteinascidia turbinata,* transparent bulbs of *Clavelina oblonga,* and the solitary black blob, *Ascidia nigra.* Occasionally, you may run across a small black anemone, *Bartholomea annulata.* Near the waterline, the barnacle *Balanus eburneus* crowds out a host of competitors. In quiet waters, the mangrove oyster, *Crassostrea rhizophorae,* may dominate most of the root system, growing in great clumps both on the roots and on one another. It is edible and harvested commercially in Puerto Rico.

Two species of fanworms attach to mangrove roots: the feather-dusters, *Sabella melanostigma,* which has red-and-white gill crowns, and *Sabellastarte magnifica,* with a large brown white-tipped ring of gill filaments. These polychaetes filter-feed by trapping food in the fine hairs of the filaments that make up the crown. Both inhabit quiet, silty places in preference to open water.

Finding a place to hide in this world of curving stalks and hostile water takes some doing. A moving sponge or clump of algae gives away the **decorator crab,** *Microphrys bicornutus,* who scavenges over the roots. Other creatures take refuge within sponges. The gray *Ircinia,* mentioned earlier, harbors a red snapping shrimp, *Synalpheus brevicarpus,* several kinds of crabs, and at least two species of brittle stars. If you can fetch out a brittle star from its hiding place, put it on a clean sand bottom and watch its fast, serpentine movement, which is nothing at all like the slow gait of the sea star. Brittle stars avoid light, venturing out only at night.

MUD FLATS

In shallows adjacent to mangroves, but as often bounded by grass or sand bottom, the fine silts of the mud flat are filled with burrowers and surface-grazers. Spotted and streaked in white-on-black or dark red, the variably patterned snail, *Neritina virginea,* browses on the film of algae that grows on the surface of the submerged mud. Soft bottoms hold so many species of snails you will need a shell guide if you are bent on identifying every one you find. Many burrow beneath the surface. Carnivorous moon snails, the small *Natica canrena* and the large **sharkeye,** *Polinices duplicatus,* plow through the silt hunting clams. The fiddler crab, *Uca,* scavenges near the water's edge. The male has one claw as big as his body. During mating season he scrapes it with his small claw in much the same fashion as drawing a bow over a violin.

The flats are pocked with holes and lumps of mud. Some are the doing of the fiddler crab, but many are the work of other animals. The sea cucumber *Holothuria arenicola* heaves up large mounds. Stomatopods and shrimp dig in so quickly that your chances of intercepting them without resorting to unfair subterfuge are slim. Biologists do it by injecting the hole with a slow-setting polyester resin that not only traps the creatures within the hole but preserves the shape of the burrow. Many holes are multichambered and inhabited by others aside from the original burrower. Side shafts occupied by the latecomers hold fish, crabs, and annelid worms.

Mud-flat crabs also burrow, even the active swimmers. *Callinectes ornatus* lives in a burrow while young but only digs in deep enough for cover when adult. The bright-red **box crab,** *Calappa flammea* also scuttles into the mud for cover, exposing only its mouth and the forward ridges of its flattened, triangular claws.

If you dig up and sift the silt, you will uncover clams, burrowing worms, anemone, brittle stars, heart urchins, sea cucumbers, and other odd creatures, depending on the fineness or coarseness of the silt. The finer sediments are poorly oxygenated but firm enough to create a stable burrow. Coarse sediments, better oxygenated but too loose for burrowers, hold those who come up for a meal and a fresh change of water at regular intervals.

Channels and Sloughs

In the deeper waters of the mud flat and the channels through which the tides drain out from the mangroves, the bottom is sure to be strewn with the **upside-down jellyfish,** *Cassiopeia xamachana*. A gelatinous pie plate with fronds, its round disk has a mouth at its center from which extends frilled tentacles colored yellow with zooxanthellae. Pick it up (with gloves; it's mildly irritating on tender skin) and give it an upward boost. It will pulse just like its pelagic cousins but quickly settles back down to the bottom, frond-side-up. If not, it flops over to expose the symbiotic algae in its fronds to the sun.

9

Man-Made Places and Predicaments

WHETHER BY DESIGN OR NEGLIGENCE, HUMAN ACTIVITY HAS placed considerable unnatural material in the sea. Wharf pilings, stone jetties, concrete sea walls, riprap, artificial reefs made from leftover scraps, oil rigs, and shipwrecks make up the bulk of the solid stuff. The sea has also become a convenient disposal site for unwanted liquids and watery solids: silts, sewage, sludge, dredge spoil, chemical waste, hot water, garbage, and oily bilge water to name a few.

The threat of inadvertent pollution of the sea grows every year with the international traffic in oil, chemicals, and radioactive material. Oil spills have, thus far, wreaked the greatest havoc, although far more damage is done by oil seeping off the land into the sea than by direct spills.

SHIPWRECKS

Along the Atlantic Seaboard, heavy weather, war, poor judgement, and bad luck have sent thousands of ships to the bottom. Near-shore shipwrecks become well known "hot spots" to fishermen. Party boats periodically visit them when school-fishing or drift-fishing is slow.

Fish life does not immediately appear on a recent wreck, but develops as time passes. A series of pioneers must first settle there and only later attract both a permanent population as well as seasonally transient visitors. To watch the process from the beginning, one must, unfortunately, start with a calamity. In May 1963, the Dutch motorship *Pinta,* inbound for New York Harbor with a cargo of lumber, was struck by the *City of Perth* in broad daylight and calm seas, and sank within an hour. She came to rest, intact, port side on the sand, in twenty-five meters of water about fifteen kilometers east-northeast from Shark River, New Jersey.

Within six months, the hull showed signs of rust and paint scaling and had acquired a veneer of organic detritus. Growth had started on wire rigging and railings. By late November, seven months after the collision, growth on the rigging was firmly established. Dense masses of the **pink-hearted hydroid,** *Tubularia crocea,* and occasional patches of the tough, wiry hydroid *Sertularella gayi* made up most of it. The starboard side of the hull, uppermost and much of it nearly horizontal, was relatively barren except for a few small beds of mussels on the forward and after sections. None of the mussels, *Mytilus edulis,* ex-

Mussels and sea stars gain a foothold on the timber cargo of the ship *Pinta*

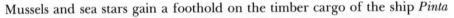

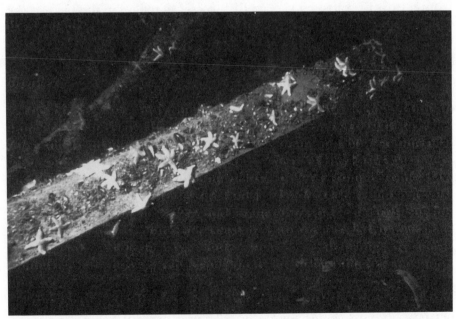

ceeded a centimeter or so in length. Every bed contained one or two tiny common **Atlantic sea stars,** *Asterias forbesi,* with arms no longer than two centimeters.

Amidships, the hull was bare, as was the deck, now vertical. The hatch covers to the cargo holds were gone, exposing the lumber, which was devoid of attached life. Except for an occasional **cunner,** *Tautogolabrus adspersus,* fish life about the wreck was nil.

By April of the following year, the mussel beds had enlarged in area and average size. Most were now three centimeters long. Their major predator, the sea star *Asterias,* had also grown. Two species were now present: *A. forbesi* and *A. vulgaris,* the **purple star.**

Mussels and sea stars were now commonplace on the masts and bridge, interspersed in and between the enlarged masses of hydroids. *T. crocea* had spread along the bridgework, the masts, and appeared in isolated patches on the hull. Its stalks had grown considerably longer on the rigging, to fifteen centimeters, swelling its apparent diameter to nearly a third of a meter. It formed a thick, tangled matrix that housed a number of species both on the stalks and at the base of the hydroids. Astonishing numbers of small (one to four milli-

Tough, wiry strands of hydroids compete for space on wreckage

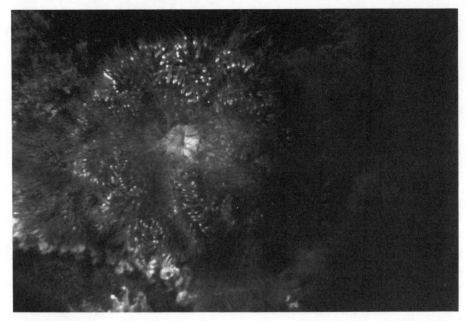

Fully expanded *Metridium senile*. Their body colors vary from off-white to dark brown.

meters) mussels were developing within them. A single grab sample, torn loose and unceremoniously stuffed into a bottle, contained over two hundred mussels, as well as nemerine worms, nereid worms, amphipods, skeleton shrimp (Caprellidae), and myriad protozoans, both attached and free-swimming.

Within the polyps, the **crowned sea slug,** *Doto coronata,* browsed on the hydroid polyps. It can ingest the cells of hydroids and transfer them, intact and unfired, into its own dorsal palps, to use for its own defense. Sea anemone covered the hull in spaces between the mussel and hydroid growth. Although varied in color—white, light orange, pink, spotted pink, dark brown, and gray—all were *Metridium senile.* This anemone has been the subject of much taxonomic confusion. Several species names have been given—*M. dianthus* and *M. marginatum,* in particular—to what now is believed to be variations of *M. senile.*

One year later, in May 1964, the sea stars were making serious inroads on the mussel beds on the hull. The sea stars were now adult

A tautog hovers over a patch of *the hydroid, Tubularia crocea*

size—arm lengths to seven to eight centimeters. Anemone were increasing in numbers rapidly, expanding into the space where the mussels were being decimated. But still the fish life around the hulk was sparse.

By November, a year and a half after the sinking, small cunner were plentiful, and a few **tautog,** *Tautoga onitis,* had appeared. **Conger eel,** *Conger oceanicus,* were living in nooks between the sand and the hull. Lobster also had found homes in holes along the sand and in the hull. On one dive, a half-dozen, one to two kilos each, were bagged.

T. crocea had disappeared from the hull but was flourishing on the bridge and cabling. The mussel beds on the flat exposed hull were decimated, the work of the sea stars and the recently arrived cunner and tautog. Anemone dominated the horizontal portions of the wreck and the hull, packed one next to the other in giant carpets.

Ten years later, the fish life around this wreck was enormous; cunner and tautog year-round, sea bass in the summer. Anemone still covered the hull, crowding out all competitors. At the edge of every sharp angle and on the borders of every opening—ports, scuppers, doorways, hatches, and the like—dense stands of hydroids persisted, as they did also on cables and railings. As in years before, they still harbored immense numbers of young mussels. Large full-grown mussels hung in tight festoons at corners, along railings, and on loose rope, but no beds of any consequence existed on flat open surfaces.

Studies of settling succession on horizontal disks done by the National Marine Fisheries Service at Sandy Hook, New Jersey, in nearby waters showed that, in midsummer, polychaete worms such as *Polydora ligni* and the hydroid *Obelia* attach initially, followed by *T. crocea*, mussel spat, sea star, and barnacle larvae. Within four months, predatory flatworms and rubber worms (Rhyncocoela) appeared. From October to December, the number of species declined and *T. crocea* became dominant.

Sea anemone often completely dominate wreck surfaces. These are *Metridium senile*.

If you visit the same location frequently, there is no reason why you cannot conduct similar experiments. Porous clay (flower pots) or wood make good substrates. Anchor them well with rope. At a later date you can cut them loose, put the pieces in a bucket, and transfer them, whole or parcel, to your aquarium.

Older shipwrecks in the Northeast often harbor growths of the **star coral,** *Astrangia danae.* It rarely grows in patches larger than your hand. This coral adapts well to an aquarium and is pleasing because it is one of the few corals in the world whose polyps open to feed in the daytime.

Fish Schools and Wrecks

In fall and winter, wrecks in northeastern waters are regularly visited by huge schools of cod and pollock. Trawlers following these fish by sonar have occasionally failed to distinguish between fish and

A pollock swims in the gloomy waters above a collapsed Texas tower one hundred miles off the coast of New Jersey

wreckage until too late, hopelessly entangling their nets in the debris. Almost every major offshore wreck along the Eastern Seaboard has net remnants somewhere on its superstructure. These nets continue to "fish" until they rot away, which can be a very long time if the netting is synthetic fiber. If you dive such wrecks, be wary of the stuff and of monofilament line left by recreational fishermen. It's all too easy to get caught yourself.

Seventy kilometers off Beaufort, North Carolina, in thirty-four meters of water lie the remains of a torpedoed World War II freighter. Her hull and superstructure are a tangled pile of steel barely six meters off the bottom. The surrounding fish life is a curious mix of temperate and tropical: **queen angelfish,** *Holacanthus ciliaris,* and sea bass swim near the bottom, side by side, staying close to the crannies formed by the collapsed debris. The attraction to the bottom-feeders who seek food and shelter around the old bones of a wreck seems straightforward enough, but upon looking up, you see another grouping that is not so easy to explain. Six meters above the highest point of the wreckage, extending horizontally nearly the length of a football field and fully nine meters thick, hang a silvery cloud of herring and scad. These fish are migratory plankton feeders, yet their visit lasted at least three days (the extent of my visit). Why stationary? Why didn't they move? During their stay, they were continually preyed upon by a circling pod of **jack crevalle,** *Caranx hippos.*

OPEN-FRAME STRUCTURES

Towers, oil rigs, and other open-frame structures below water teem with fish. Food cannot be the primary attraction: The growth on the structure couldn't supply enough food but for a fraction of the surrounding swarms. Is it cover, shade, current eddies around the tower legs, or what that draws so much life to them?

In tropic waters, jacks, bonitos, tuna, groupers, and ocean triggerfishes will cluster beneath, while barracuda and an occasional shark patrol the perimeter. In northeastern waters, open structures attract transient schooling fish. The long-collapsed U.S. Air Force Texas towers swarm with cod and pollock in the cooler months. The legs and cross-members are covered with layers of mussels in beds nearly a half-meter thick. Mussels on shipwrecks face heavy predation from sea stars and tautogs and eventually give way to other species as they are grazed down, but on open steel structures along the Northeast Coast, mussels have remained the dominant attached species for as long as these structures have been observed.

Mussels

Once a dense bed of mussels gets established, whether the common edible mussel *Mytilus edulis,* or *M. californianus,* a larger version found on the West Coast, or the **horse mussel,** *Modiolus modiolus,* the matrix formed by their shells and threads creates living quarters for a long list of other animals, and unique communities develop.

On the West Coast, *M. edulis* and *M. californianus* form mixed beds, often in association with another plankton feeder, the **goose barnacle,** *Pollicipes polymerus.* They are all subject to predation by the large sea star *Pisaster.* This association, known as the *Mytilus-Pollicipes-Pisaster* community, is a distinct, well-recognized ecological unit. Within its matrix live browsing isopods, amphipods, annelid worms, ribbonworms, nemertean worms, algae, and the scavenging crab *Pachygrapsus.*

Mussels suffer odd parasites. On the East Coast, the **mussel crab,** *Pinnotheres maculatus,* lives in the mantle cavity of the common mussel and the horse mussel, as does its West Coast counterpart, the **pea crab,** *Fabia subquadrata.* The male and female pea crab are so different in size they were once thought to be separate species. The female is the larger. Both infest the mussel just after they pass through the *megalops* stage, the last transition before adulthood. The literature often describes their relationship with the mussel as commensal, but it is not; the crab destroys tissue and absorbs nutrients that otherwise would add to the mussel's growth. Eventually the predation destroys it. On the West Coast, the gills of mussels are often parasitized by the pyncnogonoid *Achelia chelata.* For reasons unknown, the mussel never has to contend with the two parasitic species *Achelia* and *Fabia* simultaneously.

Mussels can efficiently filter nearly two liters of seawater an hour. In the summer, when vast blooms of the dinoflagellate *Gonyaulax* may appear, the mussel accumulates the toxin that *Gonyaulax* produces without harm, but then becomes poisonous to humans. The Pacific Northwest is especially prone to this problem, but sporadic outbreaks of poisoning caused by humans eating either affected mussels or clams, occurs along the Eastern Seaboard as well. Shellfishing is generally prohibited by state authorities during these "red tide" outbreaks. Should the aftermath of a feast of mussels leave your lips numb and your fingertips tingling, get to a physician.

Mussels along the Northeast Coast spawn for nearly six months, starting in summer. Their release of eggs and sperm peaks just after a full moon. When very small, just past their initial swimming stage but not yet tied down to a sedentary existence, their extensible foot, combined with well-placed extruded threads, lets them explore the bottom over a wide area. When young, a cup at the tip of their foot acts as a sucker. Their foot can extend nearly a whole body length. The combination allows them to roam over a surface looking for a good permanent site.

To see all this, as well as their cohabitants, just tear away a small patch of young mussels from their holdfasts, put them in a bottle of seawater, and bring them home with you. Keep them cool: Too much life in a small amount of water quickly lowers the oxygen supply. Cooling not only preserves the dissolved oxygen, but slows the metabolic rate. Spread the contents of the bottle out into a shallow tray and sort them out. The number and kinds of species can hardly fail to give you an appreciation for the diversity and adaptability of life in odd places as well as the formidable problem of identifying what is there and how they all contribute to the common good.

Vertical Zonation

Although vertical zonation of organisms is more often associated with intertidal life, studies of attached life on Texas and Louisiana oil rigs in the Gulf of Mexico show distinctly different species growing at specific depths. Near the sea surface, algae and the ark shell *Arca* predominate, while the oysters *Ostrea equestris* and *O. frons*, the **slipper shell** *Crepidula*, the star coral *Astrangia*, and the bryozoan *Membranipora* prefer depths of nine to twenty meters. The **American oyster,** *Crassostrea virginica*, anemones, and the barnacle *Balanus* distribute themselves randomly over the entire twenty-meter depth.

California oil platforms in thirty meters of water show similar zonation. On some rigs the **kelp scallop,** *Leptopecten latiauritus*, often crowds out everything else, existing in densities exceeding five thousand individuals per square meter, while nearby rigs may harbor none. Attached kelps do well near the surface, as do mussels. Barnacles are more prolific near the surface, but they can grow almost anywhere over the thirty-meter span.

ARTIFICIAL REEFS

Artificial reefs are created by dumping old cars, rubber tires, discarded barges, and any other locally available rubble on to a sandy or gravelly bottom in hopes of enhancing local fishing. They invariably do, but only the more substantial structures endure storm and corrosion. Some car-body reefs have utterly vanished after two seasons.

Tire reefs scatter and elicit harsh comments from fishermen who pick the tires up in their nets and from town officials who must cope with them washing ashore on beaches. On the Northeast Coast, tire reefs attract not only fish but lobster, which, in turn, attract divers. Some sessile animals will attach to rubber, but the inhospitality of that surface and constant scouring sand during winter storms keeps them bare. Reefs of concrete, brick, building rubble, broken masonry, and rock outlast and outperform those reefs made of less substantial stuff. The Schaeffer Fishing Reef off Saltaire, Long Island, was begun in 1962 with six thousand cubic meters of demolition rubble on a twenty-one-meter deep sand bottom, and still yields impressive catches of tautog, ling cod, sea bass, porgy, and flounder. Divers have spotted sculpins, butterfish, eel, ocean pout, crab, and lobster, as well as abundant attached life.

The California Department of Fish and Game compared the productivity of equal volumes of car bodies, cast concrete units, and quarry rock dumped at sea near Santa Monica. Rock and concrete units produced high numbers of fish and supported heavy kelp growth. Within five years, car bodies had corroded away and were gone.

Along the Gulf of Mexico, state agencies in Texas and Alabama have experimented with car-body reefs to enhance catches of Spanish and king mackerel, grouper, spadefish, sea bass, and red snapper. The most productive single unit along the Alabama coast is a forty-six- by ninety-two-meter concrete dry dock sunk in twenty-one meters of water. Its six-meter relief has attracted millions of fish and thousands of fishermen since it was sunk in 1957.

An enterprising diver who can relocate an underwater position that would otherwise draw little attention can, over the course of a season,

Greater amberjack, *Seriola dumerili,* patrol the stanchions of a Gulf oil rig
(Photo by D. K. Patton)

create a series of crab or lobster havens using broken pipe or water-logged crating, truck tires, or barrels weighted with stone. Connect all these objects by old rope as a guide from one to another. The real trick is to keep its location to yourself.

In the tropics, spiny lobster can be attracted to and caught by using a transposable brush pile of waterlogged stumps. After letting a heap of roots sit in a meter or so of water for a few weeks, natives drop a purse seine around it and toss the root stumps out of the enclosure to form a new pile. Lobster that sought shelter in the heap are trapped out in open sand. The technique is called "faggot farming" and is a community effort.

For the tropical skin diver, creating a few ledgelike cavities with flat rock or cement slabs is a simple enough task. Enclose the rear of the man-made cave with an easily removable rock. Spiny lobster are hard to handle from the front, but by removing the rock you can reach in the hole from the rear. Wear tough gloves and expect considerable resistance. By flexing its body, the triangular edges of the lobster's carapace open and close like scissor blades, nipping the unwary and, if in luck, saving it from a trip to the broiler.

SHIP BOTTOMS AND BARNACLES

Few are the firm surfaces unsuited to barnacles. All members of the subclass Cirripeda, these arthropods attach and prosper on ship hulls, rocky shorelines, wharf pilings, whales, lobsters, mussels, horseshoe crabs—in short, on any surface upon which they can get a foothold. There are dozens of species of acorn barnacles. Some grow a base no larger than a half-centimeter in diameter, while *Balanus nubilus,* one

Unusable culvert pipe being dumped offshore to create an artificial reef
(Photo courtesy of Alabama Department of Conservation)

Lobster hide in nooks and crannies of wrecks along the Northeast Coast

of the largest, can cover a ten-centimeter diameter and rise thirteen centimeters in height.

Their internal parts are encased by six sloping calcareous plates that rise from a cemented limey base. The central aperture is protected by four hinged plates that function as doors: open for feeding, closed for protection against predators or drying.

When open, the barnacle extends six pairs of *cirri*, feathery plumes that sweep like a cast net in the current, catching plankton. If the current shifts direction, the barnacle twists the cirri, continuously fishing into the current.

Barnacles are *hermaphroditic*, that is, each animal contains both eggs and sperm. However, they rarely self-fertilize. Clustered close to one another, they transfer sperm into their neighbor's mantle cavity with an extensible duct, which, in large species, has a reach of nearly ten centimeters.

The fertilized eggs take four months to develop into *nauplii*, the first larval swimming stage so common in plankton hauls in early spring and summer. As the nauplii molt, they transform into *cypris* larvae, a rounded two-shelled form equipped to crawl and attach to a likely surface.

And so it does after several weeks of swimming. Sinking to the bottom, the larva hangs on with its antennules. If the surface suits, it glues itself down; otherwise, it swims away and tries someplace else. Cypris larvae prefer surfaces near or upon old bases of their own species. Apparently they can chemically sense an attractant left by their kin. Arriving at a suitable site, they space themselves evenly one from another, but not from members of another species. This spacing still isn't great enough to avoid crowding as they grow. A mature bed often forces its individuals to grow in long, narrow columns, making them vulnerable to easy detachment.

WHARF PILINGS AND PIERS

Attached life develops on wharf pilings and piers much the same way it does on rocks. Cracks and crevices in the wood collect a veneer of sediment dropped by currents. Bacteria thrive in the debris. If there is light enough, diatoms grow. The surface becomes hospitable to legions of larval forms drifting in the plankton. Acorn barnacles and mussels attach. Barnacles favor the near-surface reaches to such an extent that the "balanoid zone" is clearly recognized around the world.

Over seven thousand species grow on man-made structures immersed in seawater. Six types of animals dominate: acorn barnacles, hydroids, bryozoans, ascidians, sponges, and anemone. All are plankton-feeders. At some time in their life cycles, all have a planktonic form, although most can also reproduce by asexual budding.

Hydroids, like *Obelia*, spread hollow rootlike tubes in a network over the surface, from which polyps grow on short zigzag stems. Feeding and reproductive polyps appear in such profusion that the surface takes on a furry look. When the medusae (free-swimming forms of the hydroids) mature, they break free of the reproductive polyps and pulse away. Bell-shaped, they undergo a long pelagic existence before discharging larvae that will settle down to become the next polyp stage. Once the larvae attach, things happen fast. Obelia

Surf perch take up residence among old trolley cars dumped off the California coast *(Photo by C. H. Turner)*

can metamorphose from larva to a hydroid-discharging medusa within a month.

Bryozoans gain a foothold with a few planktonic larvae, which, after transformation, spread by budding. *Bugula,* a bushy growth of boxlike individuals, looks like a miniature arborescent tree.

Ascidians, or **sea squirts,** favor clean free-flowing water, but not violent wave action. They develop from fertilized eggs into a swimming tadpole larva. It is this tadpole that links the ascidians with humans; both are in the phylum Chordata. Just how we are related is a long story and best told by a zoology text. The adults are sessile filter-feeders with shapes suggested by some of their common names—**sea peach, sea grapes, golden sea vase**—and others encrust in flattened nondescript forms.

A cold-water piling community composed of anemones, sea stars, and algae
(Photo by Alan Stewart)

Similar in shape, the sponges also move in. In the Northeast, colonies of *Leucosolenia,* the **organ pipe sponge,** cluster in tiny vaselike groups. *Haliclona* extends out in large fingerlike clumps. *Microciona,* the **red beard sponge,** appears and, like its tropical red cousins, is irritating to the touch.

Anemones sometimes prevail over entire wharf surfaces, some tiny and delicate, others large. They can crowd out all competition in some places, yet hardly hang on nearby. Why? It remains to be understood.

Wood-Borers

Not only are the surfaces of pilings and piers occupied, but so too are their interiors. Wharf pilings are treated with copper arsenicals and creosote, yet the wood-borers prevail. Time and weathering dissolve away these man-made defenses, making invasion possible.

The gribble, *Limnoria lignorum;*
and a boring amphipod, *Chelura terebrans*
(Illustration from A.F. Arnold)

A wharf piling with an hourglass shape at the water line, or a piling that shows a stump end that looks as though it had been gnawed off by a beaver is the victim of the **gribble,** *Limnoria.* It burrows just under the surface of the wood, going deeper only when the outer layer sloughs off. A swimming isopod, less than a half-centimeter in length, the gribble digests cellulose directly without aid from bacteria in its intestinal tract. Having done with one piling, it can swim on to the next. A single piling can harbor several hundred thousand. Gribble are well suited to both temperate and tropic waters. They thrive in waters whose temperatures range from 10° to 29° C, in salinities from fifteen to thirty-five parts per thousand, and oxygen from saturation to just a hair above zero: in other words, lots of places where people want to build in the water.

Deep tunnelers are mainly mollusks; *Teredo* has a worldwide distribution. The **shipworm,** *Teredo navalis,* while still a larva, attaches on and bores into wooden planking and pilings, never to emerge. The pinprick points of entrance, barely visible, give little clue to the riddled interior. Their burrows average fifteen centimeters long and never seem to overlap with adjacent burrows. Another shipworm, *Bankia setacea,* digs a one-centimeter-diameter hole up to a meter long.

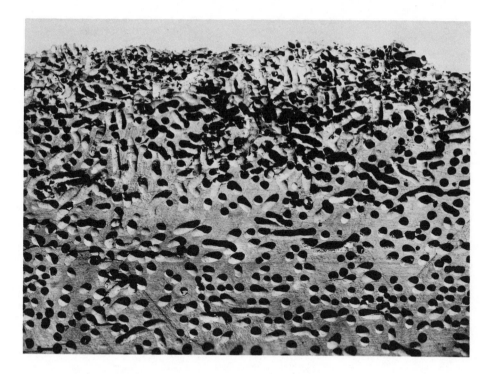

A plank totally riddled by *Teredo (Photo by Ross Nigrelli)*

Hangers-on

Floating wood that has been at sea a long time becomes the home of the **goose barnacle,** *Lepas.* Hanging on from a cemented stalk, its body is enclosed in a fleshy mantle surrounded by five thin calcified plates. Six pairs of cirri sweep the water with much the same motion as the acorn barnacle, a close relation. Unlike the acorn barnacle, whose diet is restricted to small plankton, the goose barnacle can feed on creatures nearly as large as itself.

Its reproductive life history is practically identical to that of the acorn barnacle: A hermaphrodite that fertilizes adjacent individuals (although self-fertilization is more common among goose barnacles than among acorn barnacles), it releases nauplii who develop into cypris larva that seek out new attachment sites, settle down, and metamorphose into adults. *Lepas anatifera,* which occurs worldwide, is the major fouling organism on the hulls of ships.

The goose barnacle, Lepas *anatifera,* attached to a floating coconut

MAN-MADE STRESSES

Estuaries, embayments, tidal rivers, coastal shoreline, and parts of our continental shelf are now burdened by one or more man-made stresses. Sewage in Raritan Bay, New Jersey, and outfalls off San Francisco, California, pesticide and silt runoff into Chesapeake Bay, seafood waste dumped in Conn Brown Harbor, Texas, hot water from the nuclear power plant at Chalk Point on the Patuxent River, pulp mill digest dumped off Port Angeles, Washington, sugar-cane black water choking Baou Teche, Louisiana, phosphate slime in Tampa Bay, Florida, oil-field brine in Copano Bay, Texas, and petrochemical slop in Los Angeles Harbor have all altered local underwater habitats, and not for the better.

Sewage

Aside from suspended solids, sewage carries dissolved organic matter. As it mixes with salt water, some of that organic matter begins to come out of solution. Finely suspended clays and sediments in river water entering the sea also coalesce into larger particles. The upshot of both effects is turbid water and an easily disturbed bottom sediment. At times you can see the strings of flocs in the water, like so many underwater cobwebs.

The nitrogen and other nutrients in the sewage promote the growth of algae both in the plankton and on the bottom in the vicinity of the outfall. Local effects vary; swift currents can quickly dissipate small loads and you may see little harm. But when the sewage load is heavy and the circulation sluggish, dissolved oxygen can be exhausted locally, driving away animal life. Algae may proliferate for a while and then die off, further depleting the oxygen in the water.

Nutrients, silt, and toxics from sewage and nonpoint sources (runoff) are the major causes of coastal degradation. Major estuaries, like Chesapeake Bay, have been all but destroyed by our bad habits. Since clean water is important, do what you can to work for it and support organizations who are also trying to clean up our waters.

Sludge Dumping

Some sewer *sludge,* solids settled out during sewage treatment, is still dumped or pumped into the sea. Boston and West Coast cities pump sludge. New York City hauls its sludge to sea in barges. At one time, the site was only twenty kilometers off the coast of New Jersey. Twenty-five meters deep, over the years the site bottom became a black goo of the consistency of mayonnaise. The only life that survived there were bacteria and a marine sludgeworm. The site has since been moved to one that is 160 kilometers offshore in much deeper water. Still, environmentalists worry about the effects of this effluent on marine life.

In part, pressure to move the site arose after a phenomenal blight struck the waters of the New York *Bight,* the shelf waters from the eastern end of Long Island to the southern tip of New Jersey. In early summer 1976, the dissolved oxygen in the water below the thermocline, the border between the stratified warm top layer and colder bottom water, fell to zero, destroying all the bottom-dwelling species over a sixty-five-hundred-square-kilometer area. Recreational divers noticed the first signs of stress. Bottom-dwelling fish were so sluggish they could be picked up by hand. Some, like ocean pout, hovered near the tops of wrecks—totally unnatural behavior for this fish. Ocean quahog, scallops, and lobster all suffered heavy mortalities.

Divers played a vital role in tracking the extent of the kill, describing its effects and vocalizing a demand for public action. Since that time, a diver observation program, run by the American Littoral Society and participating dive clubs, has kept tabs on this environment.

Dredge Spoils

Channels and harbors continually silt in and must be dredged periodically to keep them open. Even digging up relatively clean spoil affects life in the region dredged and in the area where the silt is dumped. In the upper Chesapeake Bay near the mouth of the Sassafras River, the density and kind of organisms in the dredged and spoil areas are significantly different from life in nearby undisturbed areas.

Near big cities, harbor bottoms can be as bad as cesspool settlings (or worse, because they are contaminated with industrial waste). These are often dumped in shallow seas and capped with sand. The fate of this material—whether the toxins and heavy metals in it will stay put or return to the water column and enter the food chain—is another unanswered question seldom asked by those responsible for the dumping.

The act of dredging can affect adjacent areas. In Redfish Bay, Texas, silt from dredge operations choked out nearby eelgrass beds. Shortly after the dredging stopped, the beds recovered.

Deep channels cut in shallows, so-called dredge and fill operations used to create shoreline for development in wetlands, creates pools of stagnant water that often lose oxygen during warm months and thus have little life in them.

Oil

Catastrophic and chronic oil pollution takes its yearly toll of sea creatures in ways both obvious and subtle. Ship spills, pipeline breaks, and oil rig blowouts dump choking layers over the sea, killing by ensnarement and suffocation.

In our haste to clean up, we have often done more harm than good. When the *Torrey Canyon* spilled oil on both the French and English coasts, the French mopped up with hay. The English tried to disperse the oil with detergents that proved lethal to many organisms in dilutions as low as 0.1 part per million. Worse, the emulsified oil settled deep into the British beach sands, destabilizing them. Within months, large tracts eroded.

Six times as much oil per year leaches off land into salt water as was spilled by the *Exxon Valdez*. This is mainly motor oil and is highly toxic to planktonic larva, fish larva, suspension feeders, and makes shellfish unfit to eat.

Hot Water

Thermal plumes from nuclear plants, coal-fired utilities, and industrial processes get mixed reviews. They can attract and sustain fish that would otherwise winter elsewhere, but a plant shutdown can send temperatures plummeting, killing fish conditioned to the milder plume temperatures. In summer, depending on locale, high temperatures can reach lethal levels. The nuclear power plant at Oyster Creek on Barnegat Bay, New Jersey, typifies these problems. Unscheduled winter shutdowns have killed truckloads of menhaden. Locals make the most of it, scooping the fish up and freezing them for summer bait and chum, but the supply usually outstrips the demand. The maximum water temperature difference between the plume and adjacent water averages 7° C.

Flow rates through these kinds of plants are massive. Oyster Creek takes in thirty to sixty cubic meters of water per second. The plant at Indian Point in the Hudson River uses 150 cubic meters per second, a volume equivalent to the flow rate of an average river.

Since temperature regulates spawning in fish, a thermal plume can induce them to discharge their eggs prematurely. Staying near the plume also upsets their natural migration pattern.

Planktonic and small life, generally anything that can pass through the ⅜-inch mesh of the plants' intake screens, is pulverized by the pumps. Life also perishes on the intake screens. Estimating this destruction can be done by sampling water before entry and after discharge and measuring its biochemical oxygen demand. This assay measures the amount of dissolved oxygen used by bacteria as they consume the decomposible organic matter dissolved in the water.

Over the summer months, when Barnegat Bay is at peak biological productivity, the Oyster Creek plant shows an average daily oxygen demand of seventy-seven hundred kilograms. To put that in perspective, a large sewage plant handling twenty-five million gallons a day will discharge a daily demand of ninety-five hundred kilograms, which most municipalities would consider too high for an enclosed bay.

FISHED OUT

Human pressure on fish and shellfish stocks has depleted many species once thought virtually inexhaustible. Inshore, oysters, scallops, and shrimp have seriously declined; offshore, yellowtail flounder, cod, pollock, haddock, and more recently swordfish and shark, are being rapidly depleted.

After the serious overharvesting along our East Coast by the huge ground and midwater trawl nets of foreign factory ships, which was halted in 1977 by the enactment of a two-hundred-mile fishing conservation zone, an even larger American fleet emerged that has carried on the carnage to the point of stock exhaustion.

Commercial fishing also accounts for the destruction of a large by-catch: non-commercial species caught in trawl nets and returned, usually dead, to the sea. In international waters, the use of enormous drift nets may be exhausting stocks of tuna, marlin, swordfish, and shark.

Recreational fishermen, can, in short spurts, also take a savage toll on some species, such as bluefish, flounder, and tuna.

As the number of humans relentlessly increases, some restraint must be imposed on those who take from the sea. Commercial fishermen oppose controls, saying the survival of the economically fittest should determine who stays in the business and who fails. Fishery management specialists point out that this kind of thinking will lead to the collapse of one stock after another, and opt instead for quotas, limited entry, and banishment of wasteful fishing methods.

Do your part to help conserve marine life, however you can, either as an individual or collectively, through marine conservation organizations. Let your congressman know how you feel on current marine issues. It is your environment to save.

10

The Underwater Naturalist

Every saltwater sailor, fisherman, diver, and beach-comber is a potential marine naturalist. The first prerequisite is emotional: an overwhelming desire to be on or in the water. Next, and only slightly more rational, is an insatiable curiosity about sea creatures and a willingness to spend endless hours watching them. Getting good at it takes the same attributes as getting good at anything: building up knowledge, honing skills, developing insight, and communicating. The first two seem obvious enough; the latter pair are a bit more elusive.

The marine sciences draw on all the major scientific disciplines for explanations. Knowledge of the sea is rapidly expanding and, at first sight, in sheer volume and complexity seems overwhelmingly forbidding. How can a latent naturalist hope to encompass it all, let alone get access to it? You cannot, nor is it necessary to try. Avail yourself of what you can about what interests you at the moment and let that build from a fragmentary hodgepodge into a coherent whole. Your acquaintance with life under water will broaden with every sea trip you make. Every new identification you garner by "looking it up" will add information not only to what you have seen but about other creatures you haven't yet laid eyes on. As this new world grows more familiar, the myriad niches these creatures fill, the ways they behave,

Free diving, properly done, requires so little exertion that
a wet suit will be welcome in all but the warmest waters.

and how they influence life around them will become apparent. You
will also discover unexplored ground where no one has yet looked or
come with a ready explanation.

The skills of the underwater naturalist include the ability to go under the water and return safely. You can develop this through courses
of instruction and by practice. If you cannot cope physically with the

stresses of scuba or free-diving, you may still be able to see a great deal by surface snorkeling. Working exclusively from the surface will present challenges. When it comes to collecting and close observation, being there is the easiest solution, but depending on your alternate abilities, it is not the only one. Nets, traps, scoops, samplers, and trawls have been used for generations and remote television has recently been added to extend observation time.

Observing is another learned skill, often unappreciated or taken for granted, but critical to gaining insight where time is limited and events can happen fast. So too is recording; it is the handmaiden of observation. Nothing will improve your ability to see as much as a self-imposed demand to accurately record what you have seen.

One way to find out whether you have your own concepts and notions straight is to disclose them to someone else. First, this forces you to mentally organize what you think you know into a coherent story. Next, if that someone else knows something about the subject, you will quickly find out whether your experiences and insights correspond. Listening to others also helps foster indirect experience.

Insight, creativity, or a half-dozen or so other vague nouns are often used to describe that mysterious sense that tells one where to look, what is unique, and why it is important. As in scientific research, the hardest step is choosing the problem. Sometimes serendipity intervenes, but don't wait for it; in order to find anything, you must be looking for something. Posing a continuous stream of questions, pursuing some and rejecting most, is where to begin.

SEEING

We look at our surroundings all our waking hours but see only a fraction of them. The input from our eyes to our brain rattles along at the rate of a billion signals a second. The brain simply has to reject the bulk of this information to avoid being completely overwhelmed. Looking is so second-nature to us that we seldom ask ourselves how well or how poorly we perceive what we see. We often miss both the content and the meaning of much of the information that reaches our senses.

What you see underwater depends on more than visibility. It depends on how well you examine what there is to be seen and how well

A small leopard shark swims among fronds of giant kelp
in a three-story-high tank at the Monterey Aquarium.

PREVIEWING THE SEA

Prior to exploring an unfamiliar region of the sea, books help immensely, but don't overlook another another source that can give you a visual sense of the underwater: your local aquarium.

For example, for divers who have never visited a kelp forest before, a half-day visit to the Monterey Bay Aquarium will do more to prepare you and your companions for that realm than a long written description.

It's also much easier to study the appearance of animals in an aquarium than in the wild, especially if trying to separate look-alike species whose differences are subtle and take time to discern.

Don't count on the animals behaving the same way in an aquarium as they do in the wild, however. Regular feeding, artificial lights, the absence of predators, the limitations of their surroundings, crowding, and forced associations all modify their lifestyles and reactions in a tank.

To find the locations and descriptions of aquaria throughout the United States, look up A. L. Pacheco and S. E. Smith, *Marine Parks and Aquaria of the United States,* 1989, Lyons and Burford, New York. It provides mailing addresses and telephone numbers as well as general descriptions that will allow you to check out what is on display before making a long trip.

you interpret it. You can be fooled by camouflage, for example, but once you know the kinds of guises creatures use, you will see through them at a glance. Most of what we miss underwater comes about by assuming subtle actions are neither significant nor noteworthy. We expect to be entertained by exaggerated and bizzare antics. But secrets are hidden in small, often brief, nuggets of action.

Understanding life underwater will start in earnest only after you quell your initial urge to see and do everything at top speed. Settle down and assimilate what one scene, one animal has to tell you. You can enhance your perceptive powers through practice. Perceiving is more than looking. Looking is passive, without analytical thought about what it is you are seeing. To *perceive* is to compare, sort out details, note relationships, find differences. What you are after is knowledge and, as the anthropologist and philosopher Gregory Bateson put it: "Knowledge is a difference that makes a difference."

Your own perceptual shortcomings will be only too obvious when you make repeated visits to the same place. You will discover animals that were there all along but that you initially overlooked. Some oversights are due to unfamiliarity, some to natural camouflage, and some to your own habit patterns.

On my first tropical snorkling trip, swimming over sandy bottom, I spent so much time looking down that I failed to see schools of needlefish and halfbeaks just under the troughs of the waves overhead. Back from another trip, I noticed while going over the photographs of fish I had taken in a coral cavern that I had completely overlooked a slipper lobster clinging to the underside of the roof of the cave. Why? Overlooking the sea surface as a reasonable place to forage was due to simple inexperience. Missing the lobster (which I really would like to have looked at close-up) was another matter. I must have looked straight at it; in the photo it was plainly in view on the roof of the cavern. Was it my fixation on the fish or unfamiliarity with the habits of that species of lobster?

Our experiences on land tell us there is more to be seen on the tops of surfaces than on their undersides. Underwater, where gravity plays a reduced role, one surface is as handy as the next and undersides are a good deal safer. It simply never occurred to me that the underside of that grotto was a place to look; therefore, I was *predisposed* to see nothing.

Because you don't see anything doesn't mean a place is bare of life. The bottom of Somes Sound in Maine, one of the few fiords on the

East Coast, is covered with loose shale. It looks utterly barren from above, yet under each loose slab, one or more brittle stars finds refuge.

Finding cryptic animals requires hunting and an awareness that, under the water, almost all spaces and cavities are put to use by something. Even then, some animals never expose themselves during daylight and retreat into such dark and inaccessible recesses that night-diving is the only way to find them. However, many hide in simple, accessible places. Empty shells, tin cans, loose rubble, under sand—are all likely hideouts.

Sea sand on open bottom looks deserted and lifeless as you swim over it, but stop anywhere at random and fan the sand with your hand and you will uncover worm tubes, sand dollars, heart urchins, clams, crabs, and an occasional fish. All these, and hosts of others, are lying low in an otherwise exposed terrain.

When you uncover something new to you or see it in passing, how much you see depends on how much you already know about it and its distinguishing features. Don't worry about what species it is if you can't place the phylum! Recognizing species, unless they are very common where you are looking, demands close attention to minute detail. Discerning critical differences can only be learned through practice. Initially, back on shore, thumbing through identification keys, trying to recall what you have just seen, you will find more often than not that you have missed salient features that distinguish one closely related species from another. Armed with that knowledge, you now know what to look for next time out.

WATCHING BEHAVIOR

The healthy animal is up and about, responding to the demands of living in ways unique to its kind. Each must solve common, basic problems—breathing, eating, excreting, resting, reproducing, thwarting enemies—or perish.

All creatures are limited by their intrinsic structures. A sponge is a primoidal pump whose cells act in loosely affiliated concert to keep the whole organism more or less working in unison. With that setup, it neither senses much nor responds much. On the other hand, a fish is far more organizationally complex and can come up with a variety

Nothing like a handout to draw them in (yellowtail snapper).

of actions we see as overt behavior. Simple animals have limited but appropriate responses to stimuli originating within themselves and from their surroundings. Complex animals, endowed with specialized structures and a much faster internal communication system, have a wider repertoire.

We sometimes attribute complex concepts to simple responses. An amoeba's cell contents flows or is firm, depending on the chemical stimulation it receives. In so doing it "advances" toward food, "engulfs" food, or, if irritated, "flees" from the scene. All these maneuvers are aspects of the same mechanism, differences in intensity of a single process.

Chemical stimulation controls the direction and beating rate of the whip- and hair-like structures of the flagellates and ciliates. This allows them to rapidly exploit information about their environment. They can move slowly or quickly, in a circular corkscrew fashion, the better to stay in a favorable location, or in a straight line, the better to move on to some place more favorable.

In lower animals, deciding what direction to take often depends on chemical sensing by two separated organs that both detect the stimulus. The flatworm will turn and swim in the direction of a water current carrying the scent of food. That and its subsequent choice of direction is found by seeking the path where the input to both sense organs is equal.

Understanding what an animal is up to begins with an appreciation for its sensory and motor apparatus and how well they are centrally coordinated. Jellyfish and anemone respond to touch and food by contraction and/or stinging. The stinging apparatus is fast and automatic; the nerve net serving the contractile apparatus is slow. Nor do they have much of a feedback system; an anemone will continue to move food into its mouth for as long as food reaches the tentacles until, full to the brim, it ejects it all. The nerve ring of the sea star is also slow and its responses are poorly organized. Thus its prey, by necessity, must be sendentary. Rarely does a sea star overcome a scallop or razor clam; both have the mobility to escape before the sea star secures its grip.

Arthropods have compound eyes and specialized limbs. Some limbs crush, tear, and manipulate; others are used for swimming and walking. The ganglia of their nervous system are more concentrated toward the head, in a rudimentary brain, which controls the integration of sensory and motor functions, but many responses are controlled by local ganglia. If a hermit crab's brain stalk is destroyed it can still walk, but, if removed from its protective shell, it cannot find its way back into it.

The eyes of the arthropod can use sunlight for navigation. They move such that they maintain a constant angle to it. Copepods respond this way. If an artificial light is placed in water, they swim around it, as do insects around a street lamp.

The more nervous control is centralized, the more elaborate is the animal's behavior. The higher invertebrates—shrimp, crab, lobster, squid, octopus—and the lower vertebrates—fish—show complex responses to stimuli. Ways of getting food, finding shelter, caring for young, schooling, migration, aggression, and socializing distinguish one species from another as much as does the form of each.

Each species and every individual can do no more than use to the fullest what has been bestowed upon it. Evolution has shaped its possessions and wired its programs. These can change only by mutation. For lower animals, behavior patterns are almost entirely innate, hardwired into the apparatus at hand, with little room for deviation. All the creature can do is do well or do poorly in the setting in which it finds itself. In higher animals, the wiring softens and the programs can undergo modification. The animal can adapt to its surroundings and increase its chances for survival. However, each generation must undergo the process anew. Learned adaptation is not genetically passed on but must be acquired either by trial-and-error or by example.

That is not to say that innate behavior mechanisms have been abandoned by higher animals. They account for much of the behavior we find puzzling in higher animals and in ourselves. The mechanisms that set these patterns have a long evolutionary history in predecessors and may now serve different purposes.

Much of the unusual behavior you will see, such as courtship rituals, depend on innate releasing mechanisms whose thresholds are set by internal changes such as hormone production and are triggered by external cues such as shapes, color, or the actions of another.

Early in life, innate releasing mechanisms answer such questions as "Who is Mom?" and "Where is home?" by *imprinting*. Salmon chemically imprint on the scent of their home stream, which will someday be their spawning grounds. To geese, Mom is the first creature seen upon emerging from the egg. These indelible stamps mark the answers to these questions for an entire lifetime.

Not all innate behavior mechanisms are necessarily helpful. Evolution has no grand purpose, no ultimate goal. Its random path, for better or for worse, proceeds without a master plan. Change for the worse is weeded out by practical trial. What may be useful or harmful later on isn't anticipated in its course. As the environment alters, what evolved before may be helpful, harmful, or irrelevant. Evolution does not necessarily eliminate what is no longer useful. Old organs may remain, like so much disconnected plumbing. Nor does evolution necessarily correct basic design flaws. The circulatory system of the reptiles has been described as a third-rate engineering job, adequate for its lethargic bearers but not at all suitable for another creature requiring speed and endurance.

The process of adaptation is a combination of genetic change and natural selection. Be careful about how you think about these adaptations. The parrotfish has fused beaklike teeth quite unlike the teeth of other fish. These teeth did not deliberately evolve to let parrotfish feed on coral polyps. Rather, fused teeth gave the parrotfish an edge on survival by giving it the means to use an unexploited food source.

Specializations such as that of the parrotfish can lead to dependence; that is, the welfare of one species becomes heavily dependent on the fate of another. This is carried to extremes in parasitism.

As you watch the behavior of individual animals, you will come to anticipate certain norms. The gill motions of a fish reflect its respiration. Its hovering or swimming posture reflects its relative state of composure. Its spontaneity and curiosity are measures of its well-being. If its surroundings degrade, if low dissolved oxygen occurs, for example, its breathing rate will increase and become labored and it will become sluggish and unresponsive.

RELATIONSHIPS

A curious result of evolution is the highly specific relationships that have developed among members of unrelated species. Cooperative or hostile, passive and unwitting, active and friendly, myriad associations have sprung up over eons. Parasitism is one. The ultimate price the parasite pays for its mode of life is specialization, degeneration, and dependence. Many imbed in a host with no more than a means to feed, reproduce, and infect another. Parasites may have little or no senses or means of locomotion. Except for their bizarre life cycles, attention to them by the naturalist is mainly a matter of their effects on their host.

A long list of worms parasitize fish, infesting major body cavities and muscles: roundworms, trematodes and cestode flatworms and flukes, and spiny-headed worms. Fluke life cycles can involve an intermediate host. "Clam Digger's Itch" and "Swimmer's Itch" are both caused by swimming larvae seeking a primary host and coming upon a human by chance.

Specialized crustaceans, copepods, and isopods parasitize a wide variety of hosts. Some attach to gills of fish, others penetrate skin. You may see them trailing like streamers from the side of an infected fish. Crabs suffer from infestation by the parasitic barnacle *Sacculina*, whose growth eventually disrupts the crab's reproductive cycle.

Deciding whether a relationship is parasitic or not isn't always clear-cut. Some snails associate with their hosts for a lifetime, yet are fully capable of independent existence. Others are so degenerate in body form that their relationship with their host is obviously parasitic.

Symbiosis and Commensalism

The word *symbiosis,* as it was originally coined, meant living together, without defining the relationship. The *symbiont* was the active partner and the *host* the passive one. Time has changed this meaning. Symbiosis now no longer includes parasitism; it is applied to mutually advantageous partnerships only.

At one time, *commensalism* meant a food-sharing partnership. Now it is commonly applied to all relationships that are neither harmful (parasitism) nor helpful (symbiosis). Many commensal relationships occur by chance. Some larva simply seek a clean hard surface upon which to settle. Often it doesn't matter whether the surface belongs to the living or the inanimate. The acorn barnacle finds the shell of a lobster as inviting as a rock. The hydroid *Hydractinia,* usually covering

Anemone are easy to find, exceptionally hardy, and keep well in an aquarium. *(Photo by Warren Egge)*

the shell occupied by a hermit crab, can get along equally well on a rock or attached to strands of *Fucus.* The sponges and anemones on the back of the **decorator crab,** *Macroeloma,* are put there by the crab. All sorts of nonsense has been written about their utility, but their selection by the crab is strictly chance. Any advantage to either is strictly fortuitous.

Intimate associations seem more numerous in the tropics. Many are species-specific. The little fish *Nomeus* is rarely found anywhere else than amidst the tentacles of the Portuguese man of war, and the conchfish *Astrapogon* within the folds of the conch *Strombus.* Corals and sponges host legions. Examples number in the thousands.

COLLECTING

No matter how many hours you spend in or on the water, some questions can only be answered with specimen in hand (better yet, in your aquarium). You may want to confirm a tentative identification or study a structure more intensively than you can in the field. Certainly the differences between a diagrammatic drawing and the real thing will quickly convince you that life is not easily represented in sketches; it is more variable and more complex than that. Take only what you need and what you can transport alive. Keep your collection containers cool and aerated. Battery-operated aerators are now available wherever live baitfish are sold.

Don't overcrowd collection containers: First, because you will suffocate or damage your catch if you do, and second, you can assimilate only so much material at one time. Watching live creatures is far more rewarding than studying their pickled remains; strive to keep everything you bring back alive and well until you have seen what you wanted and have an opportunity to return it to the sea.

If you must preserve material, you will have to use formalin, a thirty-seven percent solution of formaldehyde in water. It is noxious stuff and should be treated with respect. For most specimens, dilute one part to twelve parts of seawater. For long-term storage, add a teaspoonful of borax to every quart of diluted preservative.

Be sure your collecting is legal. Some states, notably California, regulate shore collecting, and others ban possession of certain species either because they have grown scarce (such as small striped bass in some northeastern states) or have been taken from closed waters (such as shellfish).

Most methods of capture traumatize the captive, but some methods are less stressful than others. Most fish will survive hook and line, but others will fight to exhaustion and cannot withstand the shock they will have to endure further unless you have special handling equipment.

Small fish destined for your aquarium are best trapped with a seine net or a throw net. Using a throw net requires practice. By a simple modification of the net you can catch small tropical fish hiding among coral rubble. Line the periphery of a one-meter-diameter circular throw net with weights. In its center, make a hole just wide enough to fit your arm. Lay the net over the coral, and with your arm in the net and a plastic collection bottle under the net, turn over the rubble. As the fish darts out, trap it in the folds of the netting with your free arm, then transfer it into the collection bottle.

A variation on this technique is to draw fish into the net area with bait. Two divers can spread the net and let it settle. Some fish will head for its periphery and escape, but many will duck into the nearest shelter and can be easily retrieved.

A lift net or umbrella net is also useful, especially at night from a boat. Lower it open with a light at its center, then close and raise it as fish are attracted to it. During the day, replace the light with bait.

A slurp gun, basically a wide-mouthed syringe, can be useful, but fish have a way of darting out as quickly as they are sucked in unless the barrel is fitted with a one-way flapper valve.

Aquariums

If you plan to collect, have your saltwater aquarium set up and running at least three weeks before you attempt to keep specimens. Remember, your aquarium is a far cry from a natural environment. You cannot duplicate the animals' natural setting, but you can provide similar salinity and dissolved oxygen content as well as a consistent waste level, light level, and temperature. Without going into a long how-to-do-it, let's list a few essential do's and dont's:

• Except for holding or culturing tanks, don't bother with anything under twenty gallons. For photography, the high, deep form of tank is more useful than the standard, shallow form.

- Keep tanks out of direct sunlight. They get too hot and you can't control algae growth.

- Equip your tank with a subgravel filter system and a cover. Also, an outside filter that works by airlift is useful to remove particles, reduce turbidity, and remove color if things start to go haywire.

- Usual instructions suggest a dolomite gravel bottom. If you want a sand bottom, use dolomite underneath it. If you use a two-layered bottom, either separate them with a layer of plastic fly-screen, or keep diggers out of the aquarium. They will quickly plow the sand and gravel into a hopeless mess.

- For your air supply, get a well-made pump, either a piston or vibrator type. Make sure it has more than adequate capacity and is quiet.

- Don't over-decorate your aquarium; better yet, don't decorate it at all except to provide cover for cryptic creatures. Immobile objects hamper water circulation.

- Use synthetic seawater. If you can get clean seawater, hold it in the dark in closed containers for a month before using it to eliminate the plankton it contains.

- Condition the aquarium with a few hardy fish for several weeks before adding delicate specimens. Conditioning, promoting the growth of bacteria that break down the excrement of animals, stabilizes the waste levels of nitrogen in the water.

- Don't overcrowd an aquarium. An old rule of thumb is one inch of fish for every gallon of water. Another is three inches of fish or invertebrates for every square foot of filter bed.

- Try not to add more than one new species at a time. Give each newcomer time to acclimate to the water and new surroundings.

The odds that local marine life will adapt to your aquarium depend on species, size, the creatures' condition, water temperature, and how well they survive capture and transport. The easiest acclimation from

Sea horses are easy to catch in a seine net and good aquarium additions. Feed them live brine shrimp. *(Photo by Ken Rice)*

nature to tank occurs at that time of year when the outdoor water temperature is close to that in your aquarium. Don't expect winter flounder pulled from 1° C water to make it in a 20° C aquarium.

Some species are too active for a small aquarium: young bluefish and mackerel for example. Herring and menhaden don't transfer well. They lose their scales easily unless carefully handled.

Certain invertebrates can be kept in tanks stocked with fish. Anemone, sea urchin, small hermit crab, and sea stars do well. Crabs and

If you find a skate's egg case intact, you can raise the embryo to "birth" in a glass container placed in an aquarium without other occupants.

lobsters can also be kept, but they are best housed away from potential prey. You can keep scallops, clams, and other mollusks, but don't mix them in with natural predators like sea stars or they will be quickly wiped out. Avoid predator-prey pairings; the prey have little chance.

Jellyfish, comb-jellies, many hydroids, and worms don't do well at all in an aquarium. They can be held in cool aerated seawater for a few days but rarely survive much longer.

If you can set up a small holding tank, it is often helpful to put a new specimen there for a few days prior to transferring it to a larger tank. For example, an anemone attached to an undesirable object like creosoted wood can be coaxed onto a better surface in a holding tank before it is put in the main aquarium.

PHOTOGRAPHY

The natural beauty of living creatures diminishes rapidly in death, whether you pickle them or not. Their shapes change and their colors fade. Rather than take back a ghastly remain, why not capture its living image?

Not so simple, you think. Nature's masterpieces come in small packages that even when removed from the water or transferred to a portable aquarium are tough to photograph, or are so large and so fast that you must do the job in the water. Both problems have been solved by modern thirty-five-millimeter equipment.

If you obtain a single-lens-reflex camera fitted with a macro lens, you can photograph small objects close-up and get a film image as large as natural size. New cameras have built-in exposure systems that compensate for all the lens extension complications you had to grapple with in older devices.

It is much the same story in underwater equipment. There you have additional complications: You need auxiliary light, usually

Rigged for night photography, this underwater camera is prefocused on the centered intersection of two flashlight beams, which locates the area illuminated by the electronic flash. *(Photo by Jerome Prezioso)*

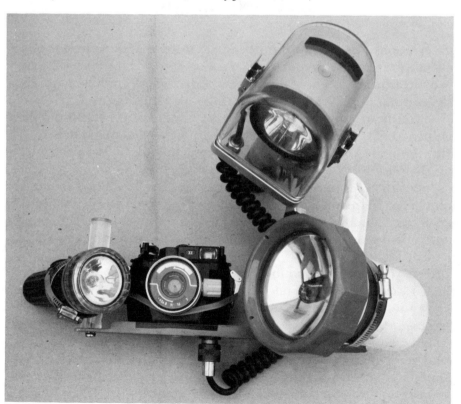

strobe. You can encase a thirty-five-millimeter camera (especially useful if you intend to do a lot of close-up work underwater), or you can get cameras especially built for underwater work. If your pocketbook can afford it, go for these systems. If not, look up older, but less expensive, ways of doing the job: use of supplementary lenses or extension tubes for close-ups and less automatic underwater rigs. Both can be taxing at first, but many fine photos were taken using those methods.

No matter what you choose, practice and keep notes on what you are doing. No sense constantly repeating the same mistakes over and over again. And sharpen up your above-water general photography. Until you know the rudiments of focusing, proper exposure, and composition, it makes little sense to plunge into the tougher aspects of the photographer's art.

KEEPING A LOG

Observations forgotten are no better than observations never made; and it's even worse if your hazy recollections slip into fiction. Accurate recording is essential if you plan to become a credible naturalist. Well-written notes are your everlasting link with what you have seen, what you have found out, and what you made of it all at the time it happened. The very process of writing it up will demand you collect your thoughts, examine what you have seen, and mull over its meaning. It will also make clear what to look for next time.

Underwater notes, if made at all, must be brief shorthand reminders to help you write up field notes once you are high and dry again. As soon as you comfortably can after a dive, jot down enough key words to jog your memory when you write up your day in a permanent log.

Keep your log in a bound, page-numbered book. Stationers call them a "Record Book." They are made of high grade paper, will take heavy use, and last a lifetime. Your notes can be written in pencil or permanent ink; both have the advantage of not smearing when wet.

Start your entry with the date, local time, location, weather, sea conditions, temperature of air and sea, visibility, and anything else you feel is pertinent. Imagine someone else reading them, trying to reconstruct the place and events you are about to describe. A sketch helps; it pinpoints your view of the surroundings.

How you write up your descriptive notes is a matter of style. You can do it in terse statements, or you can interlace your observations with feelings, interpretations, or what you have found others have written on a similar subject.

If you can, write for a reader, even though it will likely be you alone who is that reader, years hence. Let the reader know what is fact and what is conjecture. Take a species identification. If it is a guess, say so. You can add your reasons if you choose. If you don't wish to pursue it simply add a question mark in parentheses (?) to mark your uncertainty.

If you take samples or photos, reference them in your notes so that you have a one-to-one correspondence between notes, specimens, and photos. If you examine your specimens at a later time, write them up with sketches in the same log and cross-reference them back to the original entry.

Develop a faith and a will for log entries. Believe in their importance. Make the log an integral part of your efforts to know more about the marine world. Don't let your entries slip or turn into drudgery. If you do get behind, forget the past and start anew. Don't lose the zest of your original intent.

Unless you are accustomed to it, writing takes practice. Your early entries may be muddy and imprecise, but give yourself a chance. Your eye and your pen will sharpen as your perspective and your mind gain in precision.

In years to come, you will never accuse yourself of writing too much but, most surely, of writing too little. Those days resavored will always be reflected on as better days, those ventures undertaken as among the best to have come your way. Your exploits into the underwater wilderness may be the most unique events of your lifetime. Record them well. There may come a time when memories of things past become the most relevant part of things present.

Measurements

Verbal descriptions will cover a lot of ground but, in some circumstances, a few significant measurements will say volumes. Especially on seawater; temperature, visibility, salinity, and dissolved oxygen can all be easily quantified.

Temperature often changes with depth. In temperate regions, thermal stratification starts in the spring and extends well into the fall, creating an abrupt region of change over a short vertical distance: the thermocline. You can measure the difference between the top and bottom water with a combination thermometer-depth gauge available in fishing tackle supply stores. You lower it on a plumb line and it will tell you the greatest depth reached and the temperature at that depth.

As sunlight penetrates the water, it is not only absorbed by the water but scattered by particles in it. The absorption by water is quite constant irrespective of salinity or temperature, but the number and size of tiny plants, animals, and other particulates greatly changes its transparency from time to time and place to place. You can measure this vertical extinction of light with a homemade *Secchi disk:* a simple, weighted, white circular piece of plywood twenty centimeters in diameter. Lower it over the side until it just vanishes. Measure the depth at which it disappears.

Salinity is defined as the total amount, in grams, of dissolved salts in a kilogram of seawater, and is expressed in parts per thousand. Seawater contains ions of sodium, potassium, magnesium, calcium, chloride, sulfate, bicarbonate, borate, silicate and twenty or so other major inorganic constituents. The relative proportions of each major component to the others is remarkably constant in all oceans. What varies from place to place is the overall concentration of salts.

You can roughly measure salinity by measuring the specific gravity of the water with a hydrometer. Any aquarium supply store has them because they can be used to maintain the proper salinity range of a saltwater aquarium. For finer measurements in the field, a hand-held refractometer is quick and easy.

Without oxygen, animal life quickly perishes. Fish become stressed below four parts per million (ppm) and invariably die below two parts per million. The maximum amount of oxygen that seawater can contain is inversely proportional to temperature. For example, at 30° C it can hold 6.2 ppm of oxygen; at 14° C, 8.3 ppm; and at 4° C, 10.3 ppm.

Dissolved oxygen can be measured electrically or chemically. The electrical method requires a "dissolved oxygen meter." You may find that, for occasional use, the chemical method will do. Kits are available, and all you have to do is follow instructions and count drops. Analysis is based on the "Modified Winkler Method." Initially, you chemically lock the oxygen up, then measure it by the addition of a reactant until a color change occurs. It measures oxygen to the nearest ppm.

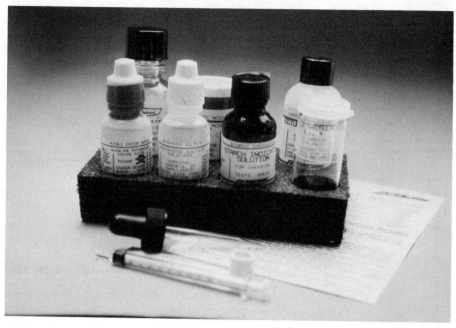

A dissolved oxygen kit contains the chemicals, apparatus, and instructions to quickly conduct an analysis of your water sample.

NOW, GET GOING

Whatever your plans, get started. Exploring under water needs no grand beginnings. Make your initial forays wherever you can find reasonably clear water and safe entry. The closer to home the better, for repeat visits will whet your curiosity, hone your ability to observe, and give you the opportunity to match what you see with what you can find out about the locale from other sources. Later on, other places will beckon. Unless you are unhampered by economic restraints, those visits will be short. To get the most out of them means going over the marine literature before you go. Your initial ventures locally will stand you in good stead. You know what will help beforehand and what must be learned on the spot, as well as what you can encompass given too little time.

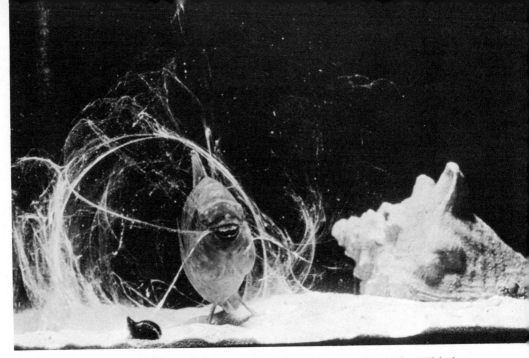

A parrotfish spins a mucous web before settling down for the night. This is much easier to see in an aquarium than in nature. *(Photo courtesy of the University of Rhode Island)*

Over the years, your visual experiences and collective wisdom about the sea will grow and provide you with satisfactions that few people will understand, not having been there for themselves.

Recall, too, that wild places and so many of the creatures that inhabit them are threatened. Do what you can to ensure that they will be there for someone else to see. Do what you can to help those who will never see them understand their value; that those animals and plants are as important to earth and sea as anything else that breathes life on this planet, including you and me.

Chapter Notes

The references given here expand on the chapter contents and will help you further explore each topic that catches your interest.

As of this writing, all the books marked with an asterisk (*) are still in print. The others are available only in libraries. Although your local library may not have a copy on their shelves, the book you seek may be somewhere in the county or state library system and your librarian can order it. However, field guides may not be loanable; they are usually assigned to the reference shelf.

Field guides are now available on sea fishes, sharks, whales, seaweeds, and reef life in general, as well as invertebrates and shore life by region. Field guides are listed here under each appropriate chapter.

Many excellent local guides exist that have not been mentioned here. They are only sold within a limited area. Check National Seashore gift shops, local aquariums, museums, and bookstores to find them.

An excellent source, albeit a trifle technical, for information on a single species, a unique community, or a specific marine region is the U.S. Fish and Wildlife Service. Write for their new publications list from:

U.S. Fish and Wildlife Service
Publications Unit
18th & C Streets, NW, Room 130–ARLSQ
Washington, DC 20240

CHAPTER 1

For more on the major divisions of life see:

*MARGULIS, L., and K. V. SCHWARTZ. 1988. *Five Kingdoms.* 2d ed. W. H. Freeman: San Francisco.

If it has been some time since you delved into biology and the invertebrates are a mystery to you, look up:

*BUCHSBAUM R., ET AL. 1987. *Animals Without Backbones.* 3d ed. University of Chicago Press: Chicago.

You may want to know more about the physical nature of the ocean and its geological history. This reference is a college text for non-scientists:

*ANIKOUCHINE, W. A., and R. W. STERNBERG. 1981 The World Ocean. 2nd ed. Englewood Cliffs, NJ: Prentice-Hall.

Metrification is coming, especially with the European community demanding that all new machinery be calibrated and constructed to metric standards. For some quick comparisons to the English system (which the English no longer use!), here are some metric milestones:

Temperature

	°CELSIUS	°FAHRENHEIT
water freezes	0	32
	10	50
	20	68
body temperature	37	99
water boils	100	212

Length

	NEARLY EQUALS
1 millimeter (mm.)	0.04 inch
1 centimeter (cm.)	0.4 inch
1 meter (m.)	39 inches or 3.3 feet
1 kilometer (km.)	0.6 mile

Area

1 hectare (ha.)	2.5 acres

Weight

1 gram (g.)	1/28 ounce
1 kilogram (kg.)	2.2 pounds
1 metric tonne	2205 pounds

Volume

1 milliliter (ml.)	1/29 fluid ounce
1 liter (l.)	1.06 quarts

CHAPTER 2

Here is a list of common field guides to fish. For tropical fish, see the notes for Chapter 7.

*CASEY, J. 1964. *Angler's Guide to Sharks of the Northeastern United States*. Highlands, NJ: American Littoral Society.

*CASTRO, J. 1983. *The Sharks of North American Waters*. College Station: Texas A&M Press.

*ESCHMEYER, W. N., E. S. HERALD, and H. HAMMANN. 1983. *Field Guide to Pacific Fishes*. Boston: Houghton Mifflin Co.

*HOESE, H. D., and R. H. MOORE. 1977. *Fishes of the Gulf of Mexico*. College Station: Texas A&M Press.

*ROBINS, C. R., G. C. RAY, and J. DOUGLAS. 1986. *Atlantic Coast Fishes*. Boston: Houghton Mifflin Co.

For general reading material on marine fishes—anatomy, physiology, and ecology—try a selection of the following:

*BULLOCH, D. K. 1986. *Marine Game Fishes of the Middle Atlantic*. Highlands, NJ: American Littoral Society.
*BIGELOW, H. B., and W. C. SCHROEDER. 1952. *Fishes of the Gulf of Maine*. Cambridge, MA: Harvard University Press.
*BUDKER, P. 1971. *The Life of Sharks*. New York: Columbia University Press.
*CURTIS, B. 1949. *The Life Story of the Fish*. New York: Dover.
*ELLIS, R. 1975. *The Book of Sharks*. New York: Grosset & Dunlap.
*MARSHALL, N. B. 1976. *The Life of Fishes*. New York: Universe Books.
*MOSS, S. 1984. *Sharks: An Introduction for the Amateur Naturalist*. Englewood Cliffs, NJ: Prentice-Hall.

Tagging fish requires equipment and know-how. For small game species contact:
American Littoral Society
Sandy Hook, Highlands, NJ 07732

For shark, contact:
NOAA/NMFS Narragansett Laboratory
South Ferry Road
Narragansett, RI 02882

For billfish and tuna contact either:
  Southeast Fisheries Center
  75 Virginia Beach Drive
  Miami, FL 33149
  or
  Southwest Fisheries Center
  PO Box 271
  La Jolla, CA 92027

CHAPTER 3

More books have been written about whales than there are right whales left in the Atlantic, according to one authority. Here is a smattering of that literature. The two volumes by Leatherwood are especially worth chasing down; enquire about their availability from the Superintendent of Documents, U.S. Government Printing Office, Washington, DC 20402.

*ELLIS, R. 1980. *The Book of Whales.* New York: Alfred A. Knopf.
*LEATHERWOOD, S., D. CALDWELL, and H. WINN. 1976. *Whales, Dolphins, and Porpoises of the Western North Atlantic.* Seattle, WA: National Marine Fisheries Service.
*LEATHERWOOD, S., ET AL. 1982. *Whales, Dolphins, and Porpoises of the Eastern North Pacific and Adjacent Arctic Waters.* Seattle, WA: National Marine Fisheries Service

Whale-watching can be done on the East Coast from:

Newburyport, MA: contact College of the Atlantic, Bar Harbor, ME 04609.
Provincetown, MA: contact The Dolphin Fleet, Box 162, Eastham, MA 02642.
Montauk, NY: contact Okeanos Foundation, Box 776, Hampton Bays, NY 11946.

For the West Coast, you can get a good guide to shore viewing:

* *Field Guide to the Gray Whale.* 1983. San Francisco: Legacy Publishing.

If you send a self-addressed stamped envelope to Legacy Publishing Company, 1850 Union Street #499, San Francisco, CA 94123, they will send you an up-to-date list of tour operators.

You can build your own hydrophone. For details and circuitry, see the "Amateur Scientist" section of *Scientific American,* October 1960, March 1964, and

August 1970, for several versions.

An excellent book on sea turtles that includes an annotated bibliography is:

REBEL, T. 1974. *Sea Turtles*. Coral Gables, FL: University of Miami Press.

CHAPTER 4

Books on plankton, especially their identification, have grown rare. Most are now out of print. The best general reading is Hardy; Newell, Smith, and Wickstead are identification guides.

HARDY, A. 1965. *The Open Sea: The World of Plankton* and *Fish and Fisheries* (combined volumes.) Boston: Houghton Mifflin.

NEWELL, G. and R. NEWELL. 1965. *Marine Plankton*. New York: Hutchinson.

*SMITH, D. 1977. *A Guide to Marine Coastal Plankton and Marine Invertebrate Larvae*. Dubuque, IA: Kendall/Hunt Publishing.

*STRICKLAND, R. *The Fertile Fjord: Plankton in Puget Sound*. Seattle: University of Washington Press.

WICKSTEAD, J. 1965. *An Introduction to the Study of Tropical Plankton*. New York: Hutchinson.

To get a plankton net, try:
Carolina Biological Supply Co.
2700 York Road
Burlington, NC 27215

Using a microscope is not as difficult as you think. The first reference will give you a quick view; the second goes a bit deeper:

*DELLY, J. G. 1989. *Photography Through the Microscope*. Rochester, NY: Eastman Kodak Company.

*GRAVE, E. 1986. *Discovering the Invisible*. Englewood Cliffs, NJ: Prentice-Hall.

CHAPTER 5

Very few books have been written exclusively on subtidal invertebrates. Most cover those found along the shores in tidal zones. Many, but not all, animals in the tidal zones are also found in deeper water.

*AMOS, W. H., and S. H. AMOS. 1985. *Atlantic and Gulf Coasts*. New York: Alfred A. Knopf.

*GOSNER, K. 1979. *Field Guide to the Atlantic Seashore*. Boston: Houghton Mifflin.

*LIPPSON, A. J., and R. L. LIPPSON. 1984. *Life in the Chesapeake Bay.* Baltimore: Johns Hopkins University Press.
*RUPPERT, E. E., and R. S. FOX. 1988. *Seashore Animals of the Southeast.* Columbia, SC: University of South Carolina Press.

CHAPTER 6

The Pacific Coast is blessed with a series of productive marine labs, aquaria, and books on coastline life. Ricketts is the guru of the relationship between the kind of coastline and its inhabitants, and well worth reading even if you never come within a thousand kilometers of the Pacific. Morris teaches as well as provides means of identification.

*BERRILL, M., and D. BERRILL. 1981. *A Sierra Club Naturalist's Guide: The North Atlantic Coast.* San Francisco: Sierra Club Books.
*CAREFOOT, T. *Pacific Seashores.* 1979. Seattle: University of Washington Press.
*KOZLOFF, E. 1983. *Seashore Life of the Northern Pacific Coast.* Seattle: University of Washington Press.
*MORRIS, R., D. ABBOTT, and E. HADERLIE. 1980. *Intertidal Invertebrates of California.* Stanford, CA: Stanford University Press.
*RICKETTS, E., ET AL. 1985. *Between Pacific Tides.* 5th ed. Stanford, CA: Stanford University Press.

CHAPTER 7

Bookstores are filled with colorful volumes on tropic reefs. Be careful buying them unless you plan to go where the photos and animals described come from. Often it's the Indo-Pacific, which isn't much help in the Caribbean or the Florida Straits.

*JAAP, W. C. 1984. *The Ecology of the South Florida Coral Reefs: A Community Profile.* Washington, D.C.: U.S. Fish and Wildlife Service.
*KAPLAN, E. H. 1982. *A Field Guide to Coral Reefs.* Boston: Houghton Mifflin.
*KAPLAN, E. H. 1988. *A Field Guide to Southeastern and Caribbean Seashores.* Boston: Houghton Mifflin.
*RANDALL, J. E. 1968. *Caribbean Reef Fishes.* Neptune, NJ: T.F.H. Publications.
*STOKES, J. 1980. *Coral Reef Fishes of the Caribbean.* London: Collins.

CHAPTER 8

*Schomer, N. S., and R. D. DREW. 1982. *An Ecological Characterization of the Lower Everglades, Florida Bay, and the Florida Keys.* Washington, D.C.: U.S. Fish and Wildlife Service.

*Zieman, J. C. 1982. *The Ecology of the Seagrasses of South Florida: A Community Profile*. Washington, D. C.: U.S. Fish and Wildlife Service.

CHAPTER 9

The zonation of life on man-made structures is obvious enough near shore but is just being recognized on offshore edifices. Similarly, shipwrecks present some interesting opportunities.

*Hay, M. E., and J. P. Sutherland. 1988. *The Ecology of Rubble Structures of the South Atlantic Bight: A Community Profile*. Washington, D.C.: U.S. Fish and Wildlife Service.

CHAPTER 10

Getting in the water can be done by all and is essential to understanding underwater life:

*Adler, H. E. 1975. *Fish Behavior: Why Fishes Do What They Do*. Neptune, NJ: T.F.H. Publications.
*Clark, John R. 1986. *Snorkeling: A Complete Guide to the Underwater Experience*. Englewood Cliffs, NJ: Prentice-Hall.

Don't go to old references on aquariums. Plastics, artificial seawater, and undergravel filters have made small (to one hundred gallons) aquarium-keeping much simpler than it was twenty years ago:

*Bower, C. E. 1983. *The Basic Marine Aquarium*. Springfield, IL: Thomas Publishing.
*Moe, M. A. 1982. *Marine Aquarium Handbook*. Marathon, FL: Norns Publishing.

For more on observing, collecting, aquarium-keeping, testing, and other actions within the scope of the marine naturalist, see:

*Bulloch, D. K. 1991. *Handbook for the Marine Naturalist*. New York: Walker & Co.
Plan to visit aquaria in areas new to you. To find them look up:

*Pacheco, A. and S. Smith 1989. *Marine Parks and Aquaria*. New York: Lyons & Burford, Publishers.

Index

Note: (page numbers in italics refer to figures.)